Nantgarw and Swansea Porcelains

Howell G. M. Edwards

Nantgarw and Swansea Porcelains

An Analytical Perspective

Springer

Howell G. M. Edwards
Department of Chemical and Forensic Sciences, Faculty of Life Sciences
University of Bradford
Bradford
UK

ISBN 978-3-030-08499-8 ISBN 978-3-319-77631-6 (eBook)
https://doi.org/10.1007/978-3-319-77631-6

Softcover re-print of the Hardcover 1st edition 2018

Printed on acid-free paper

This Springer imprint is published by the registered company Springer International Publishing AG part of Springer Nature
The registered company address is: Gewerbestrasse 11, 6330 Cham, Switzerland

Preface

The concept for this book arose from many years of research undertaken by the author into the Nantgarw and Swansea china factories, whose quality output was achieved in only a very limited time frame of perhaps only five years or so in the second decade of the nineteenth century before they ceased production finally by 1820. Decoration of existing stock was accomplished for sale by local artists at public auctions up to 1826. During this research, it became patently clear that a holistic approach was necessary to evaluate the multiple and often conflicting accounts made by early writers attempting to document the history of these two factories whose success was generated through the activities and ambition of perhaps the greatest ceramic artist, experimenter and manufacturer of his time, William Billingsley, born 1758 in Derby, died 1828 in Coalport, two esteemed sites of porcelain activity in eighteenth-century and nineteenth-century England. However, Billingsley's undoubtedly finest achievement in the manufacture of the finest porcelain ever seen was accomplished in the tiny Welsh village of Nantgarw, situated on the banks of the Glamorgan Canal which linked it with Cardiff Docks, essential for the importation of the raw materials and the export of the porcelain produced there. In a way, the phenomenal success of these two Welsh factories was their undoing as in the pursuit of porcelain perfection they were severely compromised by their inability to keep pace with demand from influential clients because of unacceptably large wastage in kiln firing amounting to approximately 90% for Nantgarw and between 70 and 80% for Swansea porcelains.

In an earlier text and companion volume, entitled *Nantgarw and Swansea Porcelains: A Scientific Reappraisal* (Springer SBM BV, Dordrecht, the Netherlands, 2017), the author drew on historical documentation, eyewitness statements and analytical papers in scientific journals to embrace a holistic assessment of the evidence in a forensic approach from which several statements were deemed to be no longer be factually correct and others were seen to be based upon misinterpreted or non-existent evidence and hearsay. This holistic analysis was a compilation of existing historical reports and of other accounts where rather different interpretations had been proposed: the crux of the matter is that, unlike several other major contemporary English or Continental porcelain manufactories, there was very little evidence remaining of recipes or formulations for the production and the empirical changes imposed thereon. Also, pattern books for both Swansea and Nantgarw did not exist, and the records of the sales through the factories or through their London agents were destroyed. Despite this, the earlier work by the author summarised the current knowledge about these two factories based on the correlation between the available scientific and historical evidence. It became apparent that much of the analytical science that had been carried out on these porcelains had been summarised without consideration of the historical evidence and vice versa. In the art world, the chemical and physical analysis of easel paintings on canvas and board panels has produced much evidential information which can be correlated with artists' palettes, preparation of pigments, ground substrate treatment, underdrawing, *pentimento* and painting procedures which feeds into the attribution of artworks to a particular artist, school or genre and has been much cited for sometimes spectacular cases in exposing fraud. It is fair to say that where expert art opinion has the support of the scientific endeavour, then all is well, but a curious dichotomy exists where the two opinions differ, and then, usually the expert art opinion is taken rather than seeking a common explanation for the discrepancies which satisfies both investigations.

In the ceramics and porcelain field generally, scientific expertise historically has been in its infancy in comparison and has not yet made major revelations about sourcing of unknown ceramic wares. There are two main reasons for this: firstly, the techniques employed have almost exclusively required that samples be removed from the specimen for analysis—and, indeed, in the earliest studies relatively large quantities of the specimen were necessary for destructive analysis. Secondly, the background information about the formulations and changes made to them during the lifetime of a factory's operation are not well documented, if at all. Hence, even when the chemical data are available from the analysis, their interpretation in terms of the raw materials used has often been superficial and therefore not very helpful in the identification and attribution of unknown pieces, and lack of chemical knowledge about the high-temperature firing processes contributes significantly to this enigma. For a painting undergoing restoration, the removal of microscopically small samples for analysis is acceptable, whereas the same situation for a perfect porcelain plate would mean that the specimen has been irretrievably damaged, thus reducing its value considerably. A porcelain specimen is also a solid heterogeneous medium, for which a microsample is not always representative of the whole bulk sample. Essentially, therefore, a porcelain analyst deals with broken pieces, known as "shards", excavated from factory waste pits or broken china where the loss of sample is not regarded as being highly significant.

In the field of chemical analysis of archaeological and art objects, the mantra that science can never prove a specimen is genuine but can expose a fake has been a working *modus operandi for* some time: however, even this is now being challenged, and there are several cases in the literature where the scientific analysis of art objects believed to be fake in museum collections has caused a reconsideration of their origin to be undertaken and hence confirming their rarity and uniqueness. This current work has necessarily taken some of the scientific aspects explored in its companion volume and has extended the scientific basis of the analytical work carried out hitherto on Nantgarw and Swansea porcelains, mainly addressing the chemistry of the raw materials used in porcelain manufacture in the eighteenth and nineteenth centuries and comparing these with the elemental and molecular data emerging from the analyses, drawing from the earliest wet chemical results up to the latest published instrumental data for scanning electron microscopy and energy-dispersive X-ray spectroscopic data. The context of the problems encountered in the manufacture of porcelain is an inherent part of this account, and therefore the appropriate historical information, where relevant, has been included and the sources and information from contemporary factories will be an important feedstock for the interpretation of these analytical data. Although all of the analytical chemical data for Nantgarw and Swansea porcelains have hitherto come from damaged pieces and shards from the factory sites, a novel feature and groundbreaking content of this volume is the first account of an analytical approach which has interrogated perfect pieces of porcelain from these factories non-destructively and noninvasively for the first time: this has facilitated the examination of the chemical composition of the porcelain bodies of rare, finished and decorated pieces of Nantgarw and Swansea china which has resulted in the recognition of a novel scientific protocol for the differentiation between these porcelains and several of their contemporaries—the future realisation of such an objective for specimens of dubious origin in collections is immediately apparent and without risk of damage to their integrity. During his many decades of collecting and handling Nantgarw and Swansea china, the author has been made aware of several well-respected collections which contain fakes and misattributed porcelain and the National Museum of Wales at Cardiff has several of these bequeathed from the Morton Nance collection which he intended to be used as an *aide memoire* and *caveat emptor* for future collectors. It is also well documented that the Samson et Cie. factory in Paris, in the mid- to late-1800s, did make "Swansea"-type porcelain, which frequently then had a factory name added in stencil or script enamels by unscrupulous vendors.

Finally, although being obviously more technically oriented than its earlier companion volume, this work should nevertheless be appreciated by those of a less technical and more historical trend because of the discursive historical factual information which forms the background to the science, particularly in the sourcing of the raw materials and the efforts made to improve empirically the quality of the final product to a demanding clientele.

Bradford, UK Howell G. M. Edwards
2018

Contents

About the Author

Howell G. M. Edwards, M.A., B.Sc., D.Phil., C.Chem., FRSC, is Emeritus Professor of Molecular Spectroscopy at the University of Bradford. He studied Chemistry at Jesus College, Oxford, and completed his doctorate there followed by a Research Fellowship at Jesus College, Cambridge. He then joined the University of Bradford as Lecturer in Structural and Inorganic Chemistry. In 2003, he received the Sir Harold Thompson Award from Elsevier Science for his international contributions to vibrational spectroscopy. He is the recipient of the Emanuel Boricky Medal for 2008/2009 from Charles University, Prague, for distinguished international contributions to analytical geochemistry and mineralogical analysis. He was awarded the Charles Mann Award from the US Federation of Analytical Chemical Spectroscopic Societies in 2011 for distinguished international work on the analytical applications of Raman spectroscopy. In his research career, he has published over 1260 papers on Raman spectroscopy and its applications and is the co-editor of six books on Raman spectroscopy and its applications to archaeology, art and forensic analysis. He has had a lifelong interest in the works of William Billingsley, especially porcelains from the Derby, Swansea and Nantgarw factories, and has published a book in sole authorship with Springer Publishing, Dordrecht, the Netherlands, on *Swansea and Nantgarw Porcelains: A Scientific Reappraisal* which appeared in 2017. In addition, he has published four monographs on William Billingsley and his porcelains, entitled: *William Billingsley: The Enigmatic Porcelain Artist, Decorator and Manufacturer; Nantgarw Porcelain: The Pursuit of Perfection; Swansea Porcelain: The Translucent Vision of Lewis Dillwyn*; and *Derby Porcelain: The Golden Years, 1780–1830.* He is Honorary Scientific Adviser to the de Brecy Trust for the scientific evaluation of artworks and paintings.

List of Tables

Chapter 1
Introduction: The Dawn of European Porcelain and the Rise of the Swansea and Nantgarw China Factories

Abstract The earliest experiments in porcelain manufacture in Europe are summarised and the attempts made to emulate the Chinese hard paste porcelains and substitution of the soft paste or hybrid versions until the secret of Chinese porcelain manufacture became known in Western Europe. Movement of artisans from factory to factory assisted in the dissemination of knowledge and transfer of expertise. The distinction between hard paste and soft paste porcelains and a listing of the factories which manufactured porcelain in the 18th and early 19th Centuries. The dawn of analytical science in the early 20th Century and a listing of the published work on Nantgarw and Swansea porcelains. The early experiments of Lewis Dillwyn, assisted by Samuel Walker at Swansea, and his successful creation of the esteemed Swansea duck-egg translucency. Recognition of the transportation problems of raw materials for the manufacture of porcelain.

Keywords Hard paste · Soft paste · Dillwyn's recipes · Swansea · Nantgarw
Early analyses · Empirical approach

The start of the 18th Century saw a concerted effort by European ceramics manufacturers to emulate the phenomenally successful influx of the Chinese export "eggshell" porcelains which had become very desirable commodities in society and whose manufacturing recipes and production technologies were still a jealously guarded secret (Pollard 2015; Freestone 1999). The first mention of Chinese porcelain was made in the annals of Marco Polo (*Il Milione*, transl., 1928), who wrote about this wondrous, novel material in the 12/13th Century and who first brought a piece of Chinese porcelain to Europe. The European majolica, faience and tin-glazed earthenwares manufactured over the following three centuries were not deemed to be competitive with the lightness, delicacy, texture and translucency of the hard paste Chinese porcelain which exhibited its applied blue and white or multi-coloured enamelled decoration to perfection.

As early as the 16th century, the manufacture of the so-called Medici porcelains (Colomban 2013; Colomban et al. 2004) was undertaken between 1575 and 1587 in the Casino di San Marco, Florence, under the patronage of Francisco de Medici I, the Grand Duke of Tuscany, as an early attempt to provide an alternative to the Chinese porcelains then sweeping Europe and in high demand by society. However,

H. G. M. Edwards, *Nantgarw and Swansea Porcelains*,
https://doi.org/10.1007/978-3-319-77631-6_1

following his death in 1587 the factory ceased production. The term "porcelain" is believed to derive from the Italian *porcellana*, meaning a cowrie shell, because of the superficial resemblance of the china to the translucent outer covering of the shell (Church, *English Porcelains*, 1894). Under the patronage of King Louis XIV from 1685 similar attempts were made to make porcelain in France, culminating in the creation of a type of porcelain in 1702 at St. Cloud by Pierre Chicaneau. However, it is generally recognised that the first successful effort in the production of a commercially viable European porcelain occurred in 1708/9 when Ehrenfried Walther von Tschirnhaus and Johann Friedrich Bottger under the patronage of Augustus II ("The Strong"), Elector of Saxony, made a breakthrough in the manufacture of porcelain at Meissen, which was rapidly followed at Sevres in the Royal Porcelain Manufactory under the Royal patronage of King Louis XV (Edwards, *Swansea and Nantgarw Porcelains: A Scientific Reappraisal*, 2017a). The early European porcelains comprised a soft paste body made of clay, powdered glass, alabaster and gypsum—known as *pate tendre* in France, *frittenporzellan* in Germany and *frita* in Spain. At this time, the attempted manufacture of porcelain in Europe in emulation of the Chinese imports was hampered by the total lack of knowledge concerning the composition of Chinese porcelain: the prime mineralogical constituent of Chinese porcelain, *kaolin*, was unknown in Europe, until this source was revealed by the Jesuit priest Francois Xavier d'Entrecolles (1664–1741) in the 1720s, who penetrated successfully the closely guarded compositional and manufacturing secrets of Chinese porcelain production (McAbe 2008). D'Entrecolles was a member of the Missionaire de la Compagnie de Jesus, the Jesuit China and Indies Mission, in Beijing: in 1722 he obtained the secrets of the manufacture and composition of Chinese porcelain during the Kangxi period from his Chinese Catholic converts at Jingdezhen and he communicated the details in a letter to Father Orry, SJ, Procurator of the Jesuit Mission to China and the Indies which is reproduced in Burton (*Porcelain, Its Arts and Manufacture*, 1906). He acquired specimens of Chinese porcelain, which were analysed by Rene Reaumur in the 1730s, which seemingly are the first ever chemical analyses carried out on porcelains: the results of these analyses were found in Reaumur's papers and archive, which comprised some 138 folio volumes on his observations of natural phenomena. Soon after, kaolin deposits were discovered in Europe, firstly near Colditz in Germany and this engendered the production of "true" porcelain simulating the Chinese imports. The archival account of Reaumur's analyses has been criticised by Nicholson (*A Dictionary of Practical and Theoretical Chemistry with its Application to the Arts and Manufactures and to the Explanation of the Phenomena of Nature*, 1808) who indicates many errors therein: at the time of Reaumur's analyses, of course, chemistry and chemical analysis were in their infancy and the concepts of chemical periodicity and of the behaviour of the elements had yet to be rationalised.

In the first few decades of the 18th Century, therefore, many other porcelain manufactories were created to fulfil an almost insatiable demand for this new ceramic material and a listing of these is given in Appendix A along with their dates of foundation and original founders: these products can be generically classified into two types—*hard paste* porcelain, the true Chinese porcelain, and *soft paste* porcelain, the European alternative version. The list in Appendix A is comprehensive but not exhaustive, and many other very small porcelain factories were established which existed for only a year or two: for example, Paris had perhaps a dozen very minor factories, usually characterised by the streets in which they were located, such as the Rue de Bondy. Many of these factories made only small quantities of porcelain which is unrecognisable today. London had five porcelain factories, four of which were located on the River Thames and one, Bow, on the River Lea: examples of porcelain items from all of these factories are known today. Also, the Comte de Brancas-Lauraguais left pre-Revolutionary France and set up his manufacturing operation in London, associated with Chelsea, but this lasted for only about two years or so. Even so, he is generally credited with establishing the first manufacture of true hard paste porcelain in England preceding the efforts of Benjamin Lund at Plymouth and William Cookworthy at Bristol.

(See Appendix A).

Historically, it is found to be rather difficult to discriminate between the types of porcelain made in these early factories and the faience or majolica from which they evolved: a very good example of this is Wedgwood, which was well known for its glazed creamware, jasperware and earthenware ceramic products but authorities still include it as a porcelain factory because of Josiah Wedgwood's ventures into bone china in the late 1790s. In the Appendix A numerical list of 60 factories therefore, approximately 21, one third, are credited with being hard paste in origin; whereas this may certainly have been more of a glazed earthenware or faience although described generically as "porcelain" by writers. There is no doubt that some of this early hard paste porcelain was extremely robust and certainly meriting the description as hard: Wilhelm Martius, writing in 1793, described Wallendorf hard paste porcelain as "so hard that even sparks are emitted by friction on steel!" Examination of the founders' names of the porcelain manufactories cited in Appendix A reveals that many were set up or strongly supported by royalty and the aristocracy, so desirable a commodity was this new material in 18th Century Europe; more detailed historical accounts also demonstrate the fluidity and transfer of expertise and skills between factories, with key personnel moving locally from one works to another—a pattern which was maintained well into the 19th Century as exemplified by the two soft paste porcelain manufactories under specific consideration here, Swansea and Nantgarw. Quite a number of these start-up factories ceased trading after a few years operation because their founders died, because of takeover by other businesses, their incorporation into other factories and an economic downturn arising from war or political upheaval such as the French Revolution in 1789, yet some survived and are still operating today: notable examples are Sevres, Royal Copenhagen, Worcester, Wedgwood and Spode. Derby still exists as Royal Crown Derby, having undergone morphing through several companies after closure of the original Derby China Works in 1848. The Caughley factory was taken over by Coalport in 1799 and the Chelsea factory by Derby in 1770.

Soft paste porcelain can be further subdivided into special variants termed, *glassy* porcelain because of the powdered flints and specifically glass cullet that it contained, *bone ash* because of the ground calcined bones used as a component in the paste recipes, *soapstone* which contains feldspar or steatite in a partial or complete replacement of the china clay and a somewhat curiously termed category called "*hybrid*" porcelain. This latter category, used by Eccles and Rackham and others (Eccles and Rackham, *Analysed Specimens of English Porcelain*, 1922; Church, *English Porcelain*, 1894) seems to be a catch-all and rather indefinite descriptor of a range of variant soft paste porcelains which includes the important category of "*bone china*", first patented it has been claimed by Josiah Spode in 1790 and still being used essentially in the same form today—hence its alternative descriptor of "*modern bone china*". The definition of porcelain and of its soft paste and hard paste varieties is fraught with misconceptions: porcelain has been defined as "a strong, vitreous, translucent ceramic, fired at a low temperature with an applied glaze fired at a higher temperature", or "a hard, shiny substance made by heating clay", or even as "a hard, fine-grained, sonorous, nonporous, translucent and white ceramic that consists of kaolin, quartz and feldspathic ores fired at a high temperature". The inconsistencies between these acceptable definitions are themselves quite revealing—especially regarding the firing temperatures required and the compositions where mentioned. A consideration of the products of the European factories listed in Appendix A will quickly reveal how inaccurate these definitions are in reality: the range of "porcelains", described there are better classified under the terms of majolica, faience, bone china, and hard paste and soft paste porcelains—which certainly do not all conform to the definitions cited above. For example, not all porcelain is translucent, the compositional formulations vary immensely and some do not contain glass frit, feldspar, bone ash, kaolin and soapstone. In addition, several texts gloss over this aspect entirely or even obfuscate the definition through incorrect statements (Jorg and Wilson, *Collection of the Rijksmuseum, Amsterdam: The Ming and Qing Dynasties*, 1997) such as:

> Chinese soft-paste porcelain, which is different from European soft-paste, originated about 1700 and became popular in the second quarter of the 18th century as part of the export assortment.

It is well known that the Chinese only manufactured *hard paste* porcelains and the European response was the *soft paste* variety because of the initial lack of knowledge in European manufactories concerning the kaolin content in the Chinese porcelains. Later, the European taste seemed to revert to soft paste porcelains for their translucency and better display of enamelled decoration.

It is easy to understand now that empirical changes made to the porcelain paste composition undertaken by manufacturers in the late 18th and early 19th century which do not seem to fit with the established *soft paste* variations classified above were to be assigned to this "*hybrid*" category. However, this category, actually, is neither informative nor discriminatory and clearly a more scientific way of better describing porcelain body variations is now required (Eccles and Rackham, *Analysed Specimens of English Porcelain*, 1922). In this respect, the suggestion (Owen 2002) that it would be better to describe porcelain body variations in terms of the ratios of their key chemical components, such as CaO: Al_2O_3: SiO_2: MgO: P_2O_5 chemically describing the ratios of lime: alumina: silica: magnesia and "phosphoric acid", is a much more scientific and accurate way of gauging the body compositions and defining the porcelain categories. For example, *glassy* porcelain is expected to be high in silicon, *bone ash* porcelain is high in phosphorus, and *soapstone* porcelain is high in magnesium content. Similarly, the use of glass frit powder or cullet will involve the detectable presence of lead arising from its incorporation into high density flint glass. Hence, a much more modern descriptive analytical science of ceramics has involved the use of ternary or quaternary diagrams (Freestone 1999; Owen 2002) to plot three-dimensionally the percentages of the major components which can materially assist in the differentiation and discrimination between the *soft paste* porcelain bodies from many factories and can be used to effectively monitor the experimental changes in body composition undertaken by a particular factory to improve its output quality. Owen has proposed (Ramsay and Ramsay 2007a, b) an octahedral diagram which effectively describes the six major components that occur in soft paste and hard paste porcelains: the six apices of this octahedron are labelled with MgO and Al_2O_3 at the axial positions and CaO, P_2O_5, SiO_2 and PbO at the equatorial positions. The octahedral concept addresses the need for a chemical compositional classification which better suits the description of the paste body: for example, it may seem rather surprising that the alkaline earths, Na_2O and K_2O, which are present in most porcelain bodies albeit in small amounts, have been omitted form this diagrammatic plot whereas lead oxide has been included despite many analyses not recognising the presence of lead at all. This situation arises because Owen's intention is to enable a reclassification of bodies, for which the relative amounts of soda and potash are not deemed to be definitive and sufficiently distinctive, whereas the presence or otherwise of glass frit will have a strong analytical indicator in the lead content. Hence, Bow first patent porcelains, the so-called "A-marked" group, are hard paste Ramsay and Ramsay (2001, 2006, 2007a,b; Ramsay et al .2001, 2004, 2006) and are dominated by the elemental oxides SiO_2, Al_2O_3 and CaO and quantitative analytical plots will thus be directed at the appropriate face of the octahedron. In contrast, the Bow second patent phosphatic porcelains comprise a significant proportion of bone ash and their data produce points within the three-dimensional octahedral body. The conclusion reached by these detailed analyses is that the label *hard paste* or *soft paste* is not as clear cut as first may be supposed as it depends upon the amount of refractory china clay present and not on the presence of any glass frit.

As will be discussed further later, the understanding of the high temperature chemical reactions and processes operating in porcelain kilns was in its infancy in the late 18th and early 19th Centuries and, therefore, many of the compositional changes to room temperature recipes undertaken at that time in porcelain manufacture were necessarily empirical and targeted to answer the fundamental question: "Is the resultant product a better porcelain, or not?". The entries made by Dillwyn (1815–1817) in his work books dating from 1815–1817 (published in 1920, Appendix B) illustrate this point very nicely when describing the results of experiments undertaken with his kiln manager, Samuel Walker, to improve the translucent body of his Swansea porcelain: his handwritten, rather terse comments such as "*… a better body was achieved …*", or "*… the best yet achieved for whiteness …*", are typical and these must be correlated with his stepwise empirical changes made in the compositional variation of mineral additives to his porcelain paste formulations. An analytical chemical interpretation of these empirical changes in his porcelain paste formulation components is made later. The Dillwyn notes described are reproduced here with several modifications in Appendix B: the codes he utilised as a shorthand notation in his work notes for each major component of the porcelain paste used in his recipes are translated in the recipes here, but several incongruities will become apparent when these are considered later in more detail. For example, the DX code in Dillwyn's preamble to his work notes and recipes is cited as "glass", which is known from other documentation to have been an integral component of the Swansea duck-egg and glassy porcelain bodies, but it is curious that nowhere in the accompanying recipes does the notation DX feature as a component of his porcelain paste – even in his written descriptors of "glass frit" as a component, where it might surely have been expected to have comprised a significant proportion of the additive!

(See Appendix B).

1.1 Porcelain Manufacture in England: Bone China

After several unsuccessful experiments, the first porcelain in England of a *glassy* variety was made at Bow in 1744/45 by Heylyn and Fye (Ramsay and Ramsay 2007a, b), followed by the start-up production at Chelsea in 1748 by Goye and Sprimont, who introduced some manufacturing ideas from France. It is not generally realised that the Bow factory first patented the use of *bone ash* as a constituent of their porcelain paste, perhaps because it was not specifically noted as such as a component in the recipe for porcelain body composition in 1749, but was rather recorded under the term "*virgin earth from animals, vegetables and fossils*" (Bemrose, *Bow, Chelsea and Derby Porcelains*, 1898), that was so described through a second patent by Fye in 1748/9 (Ramsay and Ramsay 2007a, b). Sir Arthur Church in his book on English porcelain (Church, *English Porcelain*, 1894) noted the presence in his analytical data of 17.3% of phosphoric acid in Bow porcelain, which he calculated as being equivalent to 43.8% of added bone ash in the paste, a significant presence of this component. The Chelsea and Worcester porcelain manufactories started up in the period between 1748 and 1751 and thereafter saw the establishment of factories at Spode, Derby, Coalport, Caughley, Lowestoft, Bristol and Liverpool, mainly driven by the discovery of china clay and china stone deposits in Cornwall by William Cookworthy in 1755, who founded his own factory using these materials at Plymouth shortly afterwards (Ramsamy and Ramsamy 2017). Several small porcelain factories started in France in the second quarter of the 18th Century, including Sceaux, Vincennes, Mennecy and Chantilly (Colomban 2013). At this time, a rather shadowy figure emerges in the person of the Comte de Brancas-Lauraguais, the discoverer of kaolin at Alencon in France which gave a major impetus to French porcelain manufacture, who worked at Sevres and then afterwards at Chelsea, eventually making some porcelain in England under his own name. Surviving examples of his early efforts at the manufacture of porcelain in England are now very rare and, apparently, three fine examples of his work were lost in the great fire at the ceramics exhibition in Alexandra Palace in 1873, which destroyed many fine examples of early English and Welsh porcelains (Church, *English Porcelain*, 1894).

Whilst in the context of bone ash, an interesting statement appears in the work of Dossie (*The Handmaid to the Arts*, 1758) and cited by Church (*English Porcelain*, 1894, p. 30):

> The following composition will produce wares which will have the property of true china, if they be rightly managed in the manufacture. Take of the best white sand or calcined flints finely powdered, 20 lbs, add to it of very white pearlashes, 5lbs, of bones calcined to perfect whiteness, 2lbs.

This account certainly casts doubt upon the claim of Josiah Spode in 1790–1795 almost 40 years later that he was the first to introduce a bone china body to English porcelain from the inclusion of bone ash in the porcelain paste components. Chelsea used bone ash from 1760 (Church, *English Porcelain*, 1894) and there is extant documentation for its use at Derby in 1770 (Twitchett, *Derby Porcelain*, 1980, 2002; Anderson, *Derby Porcelain*, 2000) and for the additional incorporation of bone ash and high-lead content flint glass into his paste formulation by William Billingsley at Pinxton in 1796. Yet, even today, porcelain collectors and historians still associate the emergence of bone china as synonymous with the Spode factory in the mid-1790s. These English factories generated much competition for the large amounts of Chinese porcelain still being imported via trade with the Dutch and Portuguese colonies in the Far East, but Chinese porcelain still had the commercial market edge in translucency and thinness of potting over its English competitors although the artistic decoration of English porcelain, even in the rather quaintly termed "*chinoiserie*" (or Westernised versions of typical Chinese scenes depicting pagodas, mandarin figures and small bridges) generally far surpassed artistically that of many of their Chinese analogues. The advent of a high quality "*bone china*" especially at Spode in the closing decade of the 18th Century, achieved through the addition of finely ground calcined animal bones to the porcelain composite before kiln firing, although not seemingly the first attempt to do this but evidently the most successful variant commercially, gave a distinct advantage to English fine ceramics with its capability for fine potting, lightness of texture and superior retention of a glossy, white lead oxide glaze. However, this was not a true porcelain and further experiments were ongoing into the start of the 19th Century to perfect a true porcelain, mainly at the large manufactories of Derby and Worcester, who employed teams of well-respected and acknowledged ceramic artists to finely decorate their wares (Haslem, *The Old Derby China Factory*, 1876, 1879; Edwards 2018). In France and Germany, meanwhile, the factories at Sevres and Meissen held sway in the production of some of the finest European porcelain, with Sevres benefitting financially from its Royal patronage which accounted for its commercial success certainly until the advent of the French Revolution in 1792. The war with Napoleon thereafter was a key factor in the rise of the rival English porcelain factories, since the naval blockade of French ports and the government embargo imposed against the importation of French goods meant that only relatively small quantities of Sevres porcelain could be obtained and this could be achieved only through smuggling and illegal blockade running (Edwards, *Swansea and Nantgarw Porcelains: A Scientific Reappraisal*, 2017a).

1.2 Porcelain Manufacture in Wales

In the last two decades of the 18th and the first and second decades of the 19th Century a new phenomenon appeared on the British porcelain scene in the form of William Billingsley, a renowned and highly talented porcelain artist who had achieved the pinnacle of his profession at the Derby China Works under William Duesbury I between 1785 and 1795, where he was commissioned to decorate several highly prestigious porcelain services for Royalty and the aristocracy (John, *William Billingsley*, 1968; Edwards and Denyer, *William Billingsley*, 2016; Edwards, *Swansea and Nantgarw Porcelains: A Scientific Reappraisal*, 2017a; Edwards, *Nantgarw Porcelain: The Pursuit of Perfection*, 2017b). William Billingsley, however was not content to continue merely to decorate high quality porcelain and from an early age he had the drive and ambition to create a new porcelain body that would be the most highly translucent yet achieved and which would present a fine canvas for his superb and highly respected flower painting skills. Billingsley eventually achieved the pinnacle of perfection in his porcelain manufacture at Nantgarw and an example of his finest Nantgarw porcelain, photographed in transmitted light to demonstrate its beautiful and flawless translucency, is shown in Fig. 1.1—a cylindrical spill vase, decorated with exquisite sprays of natural garden flowers in 1817–1819, probably in the London atelier of Robins and Randall and commissioned by the sole London agent for Nantgarw porcelain, John Mortlock of Oxford Street. Billingsley departed from his employment as chief decorator and head of the enamelling workshop at Derby in 1795 and after several false starts in trying to achieve the manufacture of the highest quality porcelain at Pinxton, Mansfield, Brampton-in-Torksey and Worcester, he eventually set up his own porcelain manufactory in 1811 in the small Welsh village of Nantgarw (which translates as "*rough stream*"). The location of the Nantgarw porcelain manufactory site was of especial significance, situated near the River Taff and connected to the port of Cardiff by the Glamorgan Canal and sited on the Welsh coalfield which sourced the famous anthracite steam coal of very high calorific value, which itself would later prove to be so critically important for the effective operation and firing of his high-temperature ceramic kilns (Edwards, *Swansea and Nantgarw Porcelains: A Scientific Reappraisal*, 2017a). Billingsley's initial attempt to secure British governmental financial backing for his fledgling porcelain enterprise was unsuccessful mainly because of the financial constraints imposed on the nation by the ongoing expensive war with Napoleonic France and also the war with the newly emergent United States of America. Despite this, the quality of the exemplar pieces Billingsley submitted in support of his application for funding to The Council for Trade and Plantations in September, 1814, so impressed Sir Joseph Banks, who was a porcelain connoisseur and a Member of the Council funding review commission which roughly approximated to our present-day Board of Trade, that he alerted Lewis Weston Dillwyn, the proprietor of the Cambrian Pottery, to this new venture. Sir Joseph Banks was a well-respected scientist and later encouraged Sir Henry de la Beche to found the Museum of Practical Geology in Jermyn Street, London, which became a valuable repository for quality ceramics and minerals (de la Beche et al., *Catalogue of Specimens in the Museum of Practical Geology*, 1876): this Museum later held several very important porcelain items from Nantgarw and Swansea before their eventual dispersal upon its closure later in the 1800s. Dillwyn had already conceived the idea to manufacture porcelain at his Swansea earthenware factory, which he had attempted unsuccessfully once before, in 1811. After the Council of Trade and Plantations had informed Sir John Nichol of Merthyr Mawr, an influential supporter of Billingsley and Walker and their venture at Nantgarw (Nance, *The Pottery and Porcelain of Swansea and Nantgarw*, pp. 243–245, 1942), it transpired that, through Banks' intervention, Dillwyn was able to visit Nantgarw and recruit William Billingsley and Samuel Walker to move to Swansea and to establish with him the Swansea China Works in 1814: with an existing sound financial basis, and the presence of the best acclaimed porcelain decorator and manufacturer and the most highly respected kiln master now assured, Swansea produced eventually in 1816/17 the finest, most highly translucent porcelain ever made commercially in the United Kingdom, termed "duck-egg" porcelain from its characteristic blue -green colour in transmitted light (John, *Swansea Porcelain*, 1958; Edwards, *Swansea and Nantgarw Porcelains: A Scientific Reappraisal*, 2017a; Edwards, *Swansea Porcelain: The Duck-Egg Translucent Vision of Lewis Dillwyn*, 2017c). An example of this fine Swansea duck-egg porcelain is the deep soup dish shown in Fig. 1.2, decorated locally with exotic birds by William Pollard; a photograph in transmitted light is shown in Fig. 1.3, demonstrating the exceptional translucency and characteristic duck-egg green colour of this new porcelain—note the appearance of the surface decoration when viewed through the porcelain from beneath. This was indeed a very good base for Billingsley's decorating skills, but early in 1817 he parted company with Lewis Dillwyn and he and Samuel Walker again set up in Nantgarw, this time backed financially by a local surveyor, William Weston Young, and several other local investors, where later in 1817 they did finally achieve the commercial production of the finest, most translucent and whitest soft-paste true porcelain in the world. This new Nantgarw porcelain took London society by storm and the demand from clients was insatiable: paradoxically, this created a major problem for Billingsley, Walker and their new group of local investors headed by William Weston Young in that although all the porcelain they could produce was easily sold by the major London retailer, John Mortlock of Oxford Street, the kiln wastage upon each firing of up to 90% was just not sustainable in their commercial operation. Effectively, only one saleable piece from every ten manufactured would not generate a commercial viability. Even with the large premium mark-up of several hundred percent operated by Mortlocks (a mark-up of 500% for Nantgarw has been claimed!) then applying to the sale of Nantgarw porcelain in London, mostly decorated by commission at several ateliers such as that of Thomas Robins and Martin Randall (John, *William Billinsgley*, 1968) the venture was still not commercially viable for Billingsley, Walker and their backers. An example of one of the finest and most prestigious dinner-dessert services commissioned at Nantgarw from this period is shown in Fig. 1.4; this is a dinner plate from the *Duke of Cambridge* service, which was ordered by George, the Prince Regent and later King George IV, as a wedding present for his younger brother, Adolphus, and the Princess Augusta of Hesse-Cassel in 1818 (Edwards, *Swansea*

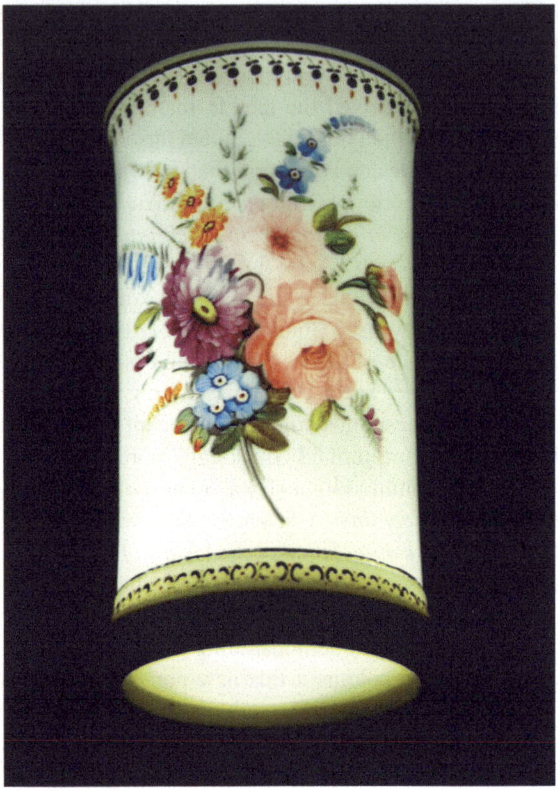

Fig. 1.1 Nantgarw porcelain spill vase, ca. 1817–1819; photographed in transmitted light to demonstrate the beautiful and clear white translucency that is so characteristic of the factory output. Private collection

and Nantgarw Porcelains: A Scientific Reappraisal, 2017a; John et al., *Nantgarw Porcelain Album*, 1975). This service was decorated by the esteemed James Plant at the atelier of Robins and Randall in London and depicts panels of fruit, birds and landscapes in the verge of a claret ground colour and a central bouquet of garden flowers. The surviving remnants of this service were dispersed in the 1889 upon the death of the then widowed Duchess of Cambridge and examples are eagerly sought after by collectors of Nantgarw and other fine historically important porcelains today.

Thus, in 1819/1820 as the results of the kiln wastage problems became insurmountable, the Nantgarw factory finally ceased production and was forced to close, whereupon William Billingsley and Samuel Walker then decamped upon invitation to join John Rose's Coalport china manufactory. No porcelain was manufactured thereafter at Nantgarw, despite some strong assertions made otherwise by several authors (commencing with Sir Arthur Church in 1894 in his *English Porcelain*), and the remaining stock "in the white" at Nantgarw was decorated proficiently and locally by Thomas Pardoe, whose premature death in 1823 resulted in the final auction sales of all surviving Nantgarw porcelain shortly afterwards. Pardoe's decorative work on Nantgarw porcelain is now rightly highly prized by collectors, although historically

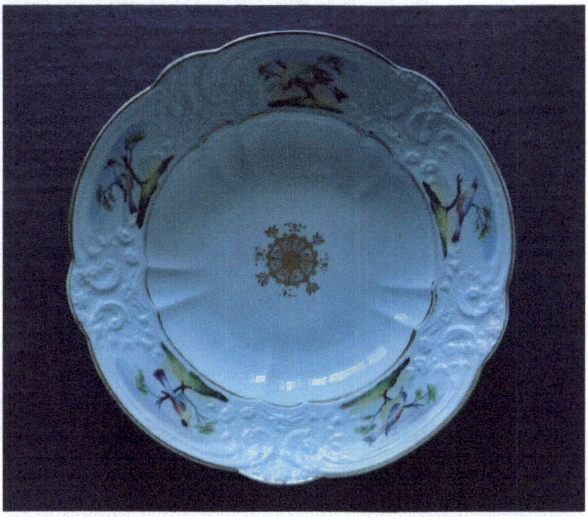

Fig. 1.2 Swansea porcelain soup dish, ca. 1817–1820; with embossed verge and decorated locally by William Pollard with exotic birds and foliage in five vignettes. Private collection

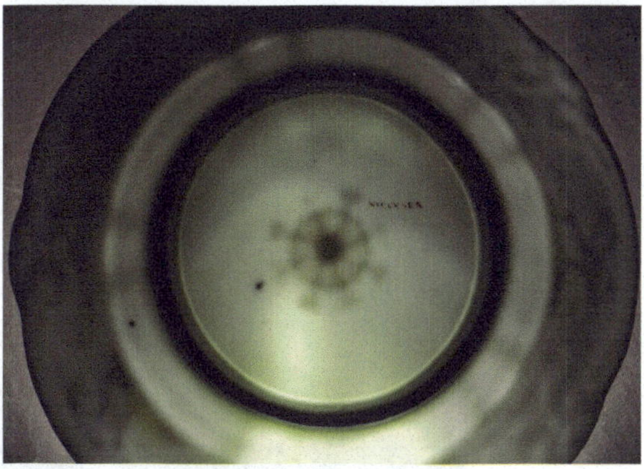

Fig. 1.3 Swansea porcelain soup dish, shown in Fig. 1.2, photographed in transmitted light to illustrate the excellent translucency, whereby the surface decoration is seen through the back and demonstrating the esteemed blue-green, duck-egg colour, so prized for the factory output. Red stencil SWANSEA mark. Private collection

he has been perhaps rather unfairly denigrated in comparison to William Billingsley (see, for example, Robert Drane in Turner, *Ceramics of Swansea and Nantgarw*, 1897). An example of the superb decorative flower painting of Thomas Pardoe is shown in Fig. 1.5, a coffee cup with a typical heart-shaped Nantgarw handle from the *Spence-Thomas* service (Edwards, *Swansea and Nantgarw Porcelains: A Scientific Reappraisal*, 2017a; John et al., *Nantgarw Porcelain Album*, 1975), painted locally after the closure of the Nantgarw factory in 1820 before Pardoe died in early 1823.

Meanwhile, at Swansea, substantial kiln losses (major, but not as serious as those experienced at Nantgarw) had also affected the production and supply of Lewis Dillwyn's duck-egg porcelain, and he took the dramatic step of trying to increase the robustness of the delicate china by replacing some of the china clay component with an increased proportion of soapstone or steatite, a magnesium silicate which also was

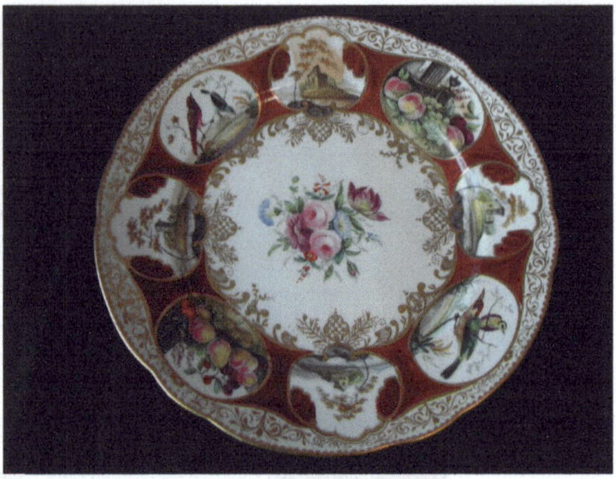

Fig. 1.4 Nantgarw porcelain dinner plate, 1818; London decorated by James Plant at Robins and Randall's atelier for Mortlock's of Oxford Street; commissioned by George, the Prince Regent, as a wedding present for his younger brother, Adolphus, Duke of Cambridge on his marriage in June, 1818, at Buckingham Palace to Princess Augusta Wilhelmina Louise, Landgravine of Hesse-Cassel. Marked NANT-GARW C.W. Private collection

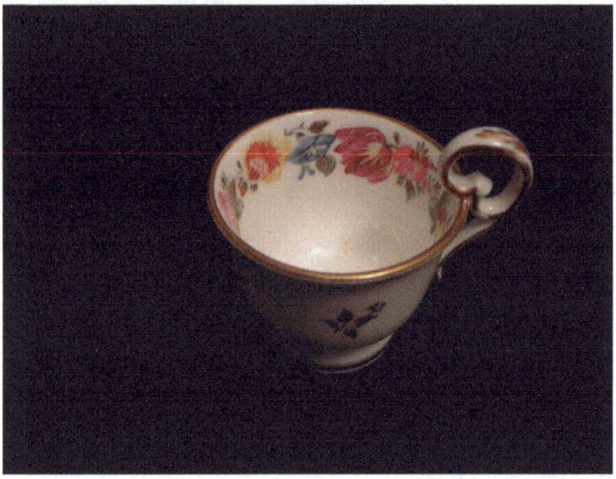

Fig. 1.5 Nantgarw porcelain coffee cup with heart-shaped handle, decorated with a floral wreath by Thomas Pardoe, ca. 1820–1823, belonging to the Spence-Thomas service. Private collection

used in Chinese hard paste porcelain under the name *petuntse*. Unfortunately, this new and more robust Swansea "trident" porcelain (so-called because of the impressed trident mark), although beautifully decorated by local artists such as David Evans and Henry Morris, did not appeal to the London clientele because of its markedly poor, rather muddy, brownish translucency and hard, pitted, pigskin -like texture (John, *Swansea Porcelain*, 1958) in comparison with the expectations revealed from the duck-egg porcelain translucency. Nevertheless, the decoration on the Swansea

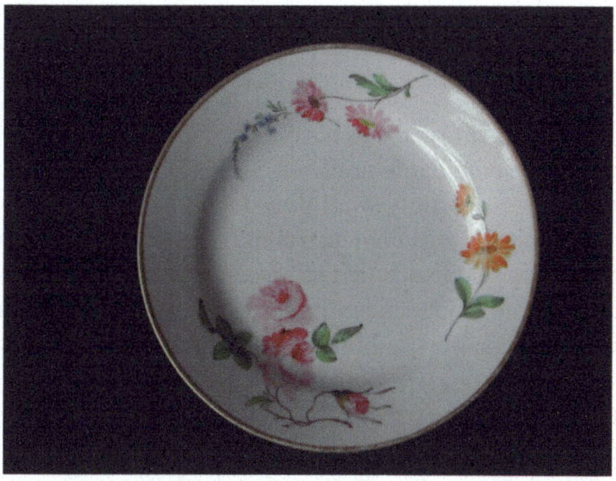

Fig. 1.6 Swansea "trident" porcelain dessert plate, impressed SWANSEA and impressed with trident motif, decorated with garden flowers by David Evans, ca. 1818. Private collection

trident porcelain is still beautifully executed and an example of a dessert plate, simply but beautifully decorated by David Evans with garden flowers, is shown in Fig. 1.6.

Hence, because of the change in his porcelain recipe, essential though it seemed for commercial survival, Dillwyn's operation at Swansea lost its market edge very rapidly with a discerning clientele and he too was forced to close early in 1820, selling out to his managers, Timothy and John Bevington. The Bevingtons then arranged for the enamelling and decoration to be effected locally on the remaining stock of glazed porcelain at Swansea to be offered for sale in the same auctions used for the residual Nantgarw stock which had been decorated for sale by Thomas Pardoe. Several authors have alleged that the Bevingtons continued to make quantities of porcelain at Swansea before the final closure of their operation but there is little evidence for this assertion and, indeed, others have stated that John Rose of Coalport removed the kilns and moulds from Swansea very soon afterwards by purchasing these at one of the first auction sales in 1823. The *Gibbins* service, of which only a few items now remain, has one piece which bears an impressed mark of BEVINGTON & SON as well as a red enamelled and stencilled SWANSEA mark—it has a duck-egg translucency, but also a rather coarse, gritty texture reminiscent of an unsatisfactory and discarded Dillwyn trial piece, and although it is suggestive that perhaps some firing of porcelain may have in fact occurred in the Bevington period then it is now believed that this cannot have been a significant undertaking (Nance, *The Pottery and Porcelain of Swansea and Nantgarw*, 1942). The idea that John Rose, proprietor of the Coalport China Works removed the moulds and also acquired the Swansea and Nantgarw recipes for his own production of porcelain in the 1820s at Coalport is a long-held belief and the present author has traced the origin of this information to an early text by William Chaffers (Chaffers, *Marks and Monograms on Pottery and Porcelain with Historical Notices of Each Manufactory*, 1863), who claimed:

> His (sic. John Rose's) rapid success (in porcelain manufacture at Coalport from 1795) enabled him to buy the Caughley manufactory in 1799, the Nantgarw porcelain manufactory in 1819 and the Swansea porcelain manufactory with their repertory of moulds. He employed William

> Billingsley, formerly of Nantgarw, as chief painter and Billingsley's chemist, Walker, who initiated at Coalport a maroon glaze and brought the Nantgarw technical recipes to Rose at Coalport.

This statement seems to have some bearing of truth in that Rose may have certainly acquired some hardware from Swansea (and also perhaps Nantgarw?), but Samuel Walker was certainly not Billingsley's chemist and the extent of Billingsley's painting at Coalport is still conjectural as, elsewhere, he is claimed to have been employed there in an advisory and overseeing capacity only in their enamelling workshop. There is no evidence at all that Rose used the Nantgarw moulds or porcelain body recipe at Coalport, although Coalport plates from the early 1820s do have an adopted style of Nantgarw moulded edge on some pieces and they do possess some superb flower painting (see for example, Fig. 1.7). A detailed comparative inspection of the Coalport and Nantgarw mouldings from this period reveals that there are several subtle differences, indicative that the Nantgarw moulds were not used by Coalport even if the general design theme was adopted there and at other factories. Chaffers goes on to claim that a new mark was then publicised on Coalport china involving an intertwined C, S and N in an ampersand (Fig. 1.8), standing for Caughley, Swansea and Nantgarw, alongside the Coalport monogram C&S, for Coalport & Salopian. Here, some dubious chronology is now evident since Godden (*Guide to European Porcelain*, 1993) ascribes the first appearance of this particular Coalport mark to the period 1861–1875, which is a period of some years following the "acquisition" of Nantgarw by John Rose and after the death of John Rose in 1848. This itself raises the question as to the commercial reason for the highlighting of Coalport's accredited acquisition of the Caughley, Swansea and Nantgarw factories and their "incorporation" into the Coalport China Works some 40 years after the event. Technically, of course, the claim that the Coalport China Works actually had "*acquired*" the Swansea and Nantgarw factories is open to further debate since both factories had already closed down prior to the auction sales in 1823 and 1824 and were never again opened for porcelain production afterwards, but the seed was sown and probably did no harm at all to the marketing claim that Coalport had taken over the operations of both of the esteemed Welsh factories in the manufacture of the highest quality translucent porcelains!

There is no doubt that, directly or indirectly, unacceptable large kiln wastage upon firing contributed to the demise of both the Swansea and Nantgarw porcelain manufactories: the exceptionally high values quoted, of between 75 and 90% of output lost at the firing stage, were just unsustainable commercially. Other factories do not mention comparable figures, but Exley (*The Pinxton China Factory*, 1963) has cited a documented example from the private papers of William Billingsley: in a letter to John Coke, principal of the Pinxton china factory, on the 22nd August, 1795. In this correspondence, Billingsley quoted some general calculations he had made relating to the setting up and production costs for china at the newly established Pinxton China Factory. Billingsley reckoned that for an estimated production run of 70 tea sets per week at Pinxton, he should allow a 1/7th kiln wastage, i.e. about 15% of the kiln firing charge—also, he referred to the "*hidden costs*" of porcelain manufacture,

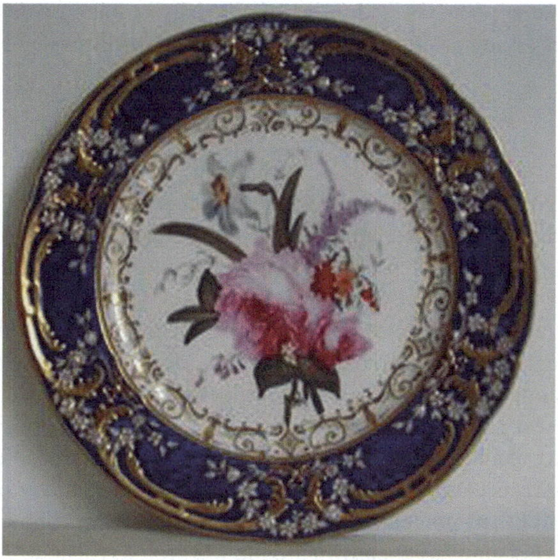

Fig. 1.7 Coalport dessert plate with Nantgarw type moulded border beautifully decorated with central floral bouquet, cobalt blue rim and heavily gilded to accentuate the border moulding, ca. 1820

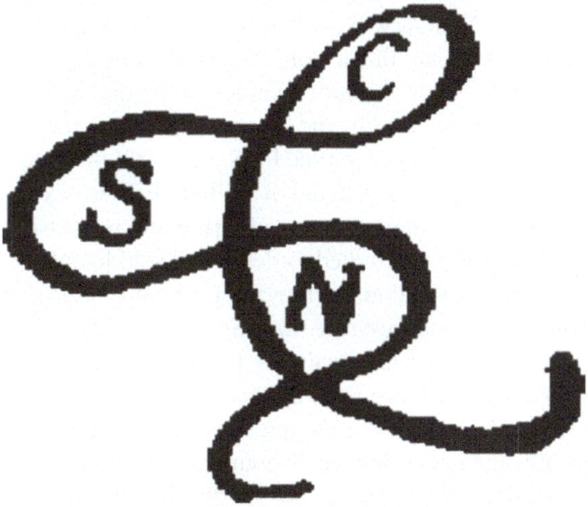

Fig. 1.8 Ampersand used by the Coalport China Works ca. 1860 and thereafter containing the letters C, S and N, claimed to represent the takeover of the Caughley, Swansea and Nantgarw factories by John Rose in 1790 and 1820–1823, respectively. Used by John Rose's successors after his death in 1848

these being essentially the kiln thermal firing costs and the transportation of wood and coal, the wood being used as charcoal for the lower temperature glost or glaze firing kiln and the coal for the higher temperature porcelain biscuit or unglazed porcelain manufacture. Some 312 tons of coal were required for kiln firing at Pinxton per annum and 510 tons of wood, costing £156 and £140 per annum, respectively, delivered to the site. These "hidden costs" referred to here were quite substantial: the cost of raw materials for the preparation of the porcelain paste being estimated by Billingsley at £23 per week, whilst the firing costs for wood and coal for the kilns were calculated at £11 per week, almost 50% of the acquisition costs of the porcelain raw materials, some of which had to be procured from some distance from the factory site. The resulting profit margins were narrow and Billingsley was moved to say to Coke:

> I am certain that it is far more advantageous to finish the ware than dispose of it in the white – dessert services I believe pay exceedingly well!

Even at this early stage in his porcelain manufacturing career, it is apparent that William Billingsley recognised the economic virtue of decorating his china on site at the works—after all, he was undoubtedly one of the most accomplished decorators of ceramics then based in Great Britain. This thought persisted to a large extent at Swansea, yet at Nantgarw the opposite strangely seemed to apply and much Nantgarw stock was sold on by Billingsley in the white to John Mortlock's retail agency in London, glazed but "in the white" for decoration in the capital's ateliers to private commissions received by Mortlocks. Larger factories such as Derby did have their own London showrooms, and indeed Derby opened a further showroom in Bath to accommodate the inspection of their already decorated and finished products by clients there during the social season (Anderson, *Derby Porcelain*, 2000), but almost all Derby porcelain was decorated on receipt of commission at their own enamelling workshops by their own artistic workforce (Haslem, *The Old Derby China Factory*, 1876; Edwards, *Derby Porcelain: The Golden Years, 1770–1830*, 2018). Today, collectors of Nantgarw porcelain in particular are divided in opinion as to whether or not the two main themes of London decoration or local decoration are comparable equally and some favour examples of the highly prestigious London-decorated services commissioned by the aristocracy whilst others favour generally simpler examples of beautifully decorated local pieces. Of the latter, those locally decorated pieces which bear the artistic hand of William Billingsley or Thomas Pardoe are especially in demand for their undisputed virtue and possession of the quadruple provenance of being made, decorated, ordered and sold locally in Nantgarw. Nantgarw services decorated in London, such as the *Duke of Cambridge*, *Mackintosh*, *Brace, Kenyon, Wilde, Duke of Newcastle, Earl Spencer* and *Lady Seaton* services can be compared with those decorated locally by Billingsley and Pardoe, such as the *Prince of Wales, Greenmeadow, Ferguson, Spence-Thomas, Ewenny Priory, Edwards, Twyning* and *Duncombe* services: a comprehensive list of Nantgarw named services decorated locally and in London can be found in Edwards (*Swansea and Nantgarw Porcelains: A Scientific Reappraisal*, 2017a) and many exemplars are illustrated in John et al. (*Nantgarw Porcelain Album*, 1975).

By 1825 all Swansea and Nantgarw china had been dispersed at auction and the manufacturing and decorating story ends at this point: however, some of the world's finest porcelain had undoubtedly been created through the efforts of the enigmatic William Billingsley who died a pauper in Coalport just three years later in 1828, having been predeceased by all members of his family, namely, his wife Sarah and their three children (Edwards, *Swansea and Nantgarw Porcelains: A Scientific Reappraisal*, 2017a). He will always be remembered for his bold and successful attempts in adverse circumstances to "*achieve the impossible*" and his legacy of Welsh porcelain is testament to the eventual realisation and fulfilment of his ambition. In this analytical case study, we shall review the data from the reported chemical determinations of porcelain body compositions carried out on the Welsh porcelains of Swansea and Nantgarw and to compare their findings: this has only been partially accomplished in the scientific literature to date, and as a result several potential misconceptions have been inadvertently promoted and accepted as factual forensic analytical evidence for these superb ceramic works of art.

1.3 Porcelain Body Variations

It can be seen from the preceding discussion that the manufacture of porcelain in the early 19th century was largely empirical and experimental, in which several modifications to the formulations and recipes were tried to achieve a better body after firing, and this often requires a cautious interpretation of analytical chemical compositional data, especially when these have been derived from just a limited range of specimens available for analysis. At Swansea, Lewis Dillwyn's diaries (Nance, *The Pottery and Porcelain of Swansea and Nantgarw*, 1942; Eccles and Rackham, *Analysed Specimens of English Porcelain*, 1922; Edwards, *Swansea and Nantgarw Porcelains: A Scientific Reappraisal*, 2017a) give a shorthand account of the experiments he carried out with Samuel Walker during the period 1816–1817, just before Samuel Walker left Swansea to accompany William Billingsley to start up again at Nantgarw, on the empirical variations in body and accompanying glaze compositions, comprising in total about twelve of the former and three of the latter (See Appendix B). At this time, it should be noted that, unlike the Swansea formulations, no lead glass frit appears in the only recipe that exists from Billingsley and Walker's previous efforts at Nantgarw and it must be conjectured that perhaps they had finally achieved the optimum composition for their paste there which was then retained unchanged over such a brief manufacturing episode?

It is not generally realised that Lewis Dillwyn had a considerable chemical knowledge, he was a Fellow of the Royal Society, which must have stood him in good stead for these empirical experiments he carried out on the formulation recipes for his porcelain bodies at Swansea even though the complex chemistry of interactive mineral reactions at elevated temperatures was still to be defined scientifically through modern analytical methods (Edwards 2015a,b); this was, in fact noted, during the legal Chancery procedures at the closure of the Swansea factory in 1820 and has been thereafter highlighted by Shaw (*The Chemistry of the Several Natural and Artificial Heterogeneous Compounds Used in Manufacturing Porcelain, Glass and Pottery*, 1837) some two decades later (Nance, *The Pottery and Porcelain of Swansea and Nantgarw*, 1942, p. 457). Shaw has been held by later ceramic historians as the first person with chemical knowledge to attempt to understand ceramic production in terms of the quantitation and chemical reactions required at elevated temperatures but, in reality, an inspection of the Dillwyn notebooks reveals the scientifically methodical way in which he approached empirically his variations in paste composition to achieve better porcelain bodies some twenty years before Shaw came upon the scene. For example, the comparative changes in soaprock and china clay quantities affected between the groups of experiments numbered in Dillwyn's notebooks as 1, 2, 4, 7 and 11, and 3, 5, 6, 9, and 10, illustrate his careful methodology and scientific approach very satisfactorily (Appendix B). Coupled with his experiments in this synthetic compositional variation, Dillwyn was in collaboration with the best contemporary kiln master in Britain in the form of Samuel Walker at Swansea, whose knowledge of the practicalities of kiln construction and the idiosyncrasies and fine tuning of the ceramics firing process were unsurpassed at that time. It can be no surprise, therefore, that Dillwyn's practical comments, although rather terse and simplistically factual, such as "*a better body*", "*improvement*" or "*the best yet achieved …*" relating to the body and the trial glaze applications he carried out must be taken to mean that Dillwyn and Walker well understood the effects of their systematic changes in paste composition upon the quality of the resultant porcelain body effected during these experiments even if the complex chemistry of the thermal processes were then still unknown. Hence, it appears that we should re-examine Simeon Shaw's thesis that Dillwyn did not appreciate the chemistry of his compositional changes "*cum grano salis*" and it is realistic perhaps that we should in future consider a case for the re-instatement of Lewis Weston Dillwyn as a major force in synthetic porcelain experimentation in the early nineteenth century.

A statement in Dillwyn's notebook, referring to one of his experimental compositions as "*... a variant of the Nantgarw body…*" has, nevertheless, created some conjecture and confusion in earlier literature as several authors have seized upon this naturally to imply that Dillwyn must have been conversant with the Nantgarw porcelain body composition—which was, nevertheless, adamantly stated to have been a closely guarded recipe (Edwards, *Swansea and Nantgarw Porcelains: A Scientific Reappraisal*, 2017a). However, it must not be forgotten that Samuel Walker, Billingsley's co-founder of the Nantgarw china works in 1813, his kiln master and his son-in-law (he was married to Sarah Billingsley) would have surely had detailed knowledge of the final successful Nantgarw paste which he could have introduced to Dillwyn's experiments in some form, whilst still preserving the secrecy of the precise compositional details of the Nantgarw formulation. If subsequent documentation is to be believed then Walker did personally finally reveal the precise Nantgarw formulation recipe to Taylor (*The Complete Practical Potter*, 1847) some years later after the death of William Billingsley, before he set up the Temperance Hill Pottery in the USA, and this formed the basis for Professor J. W. Mellor's first successful re-creation of the Nantgarw porcelain body in 1885! What it does not mean, however, is that Dillwyn used an identical formulation to the Nantgarw porcelain body at Swansea during the employment there of Billingsley and Walker: as the analytical data will demonstrate later, there are subtle analytical changes in composition between the Nantgarw body and the best Swansea duck-egg body, not the least of these being the use of lead or flint glass frit at Swansea as well as the usage of special chemical additives noted by Dillwyn in his recipes, such as smalt, arsenic oxide and borax, which are not recorded for the putative Nantgarw formulation used by Professor Mellor. Another result of the compositional changes in the relative proportions of china clay, bone ash and soapstone effected to the Swansea paste in the 1816–1819 period is the depth of colour of the duck-egg translucent body, which could vary from a deep green-blue through to a paler greenish-yellow; examples of this can be seen in the Swansea soup dish shown in Fig. 1.3, photographed in transmitted light, clearly showing the decoration on the front surface and the SWANSEA red stencil mark—this plate is the same as that shown in Fig. 1.2. Another example is the rare trumpet-shaped Swansea spill vase shown in Fig. 1.9 with a *chinoiserie* figure of a dancing Chinaman surrounded by tendrils of gilt seaweed, in a style that is very characteristic of the decoration of William Billingsley (Edwards, *Swansea Porcelain: The Translucent Duck-Egg Vision of Lewis Dillwyn*, 2017c), and a photograph of the same piece in transmitted light exhibiting a deeper blue-green duck-egg colour, as seen in Fig. 1.10. The depth of colour of this spill vase should be compared with that of the Swansea plate in Fig. 1.3. Although not specified by Dillwyn, Nance (*The Pottery and Porcelain of Swansea and Nantgarw*, 1942) has commented that a series of experiments carried out by the Ceramics Society in 1916 on synthetic Swansea porcelain made according to Dillwyn's recipes showed the greatest translucency and depth of blue-green colour was achieved near a 40% bone ash composition with equal proportions of china clay and soapstone in the formulation. When the china clay content was increased relative to the soapstone a discolouration to a peach-yellow became more pronounced and when the soapstone proportion was increased and that of the china clay decreased the blue colour disappeared altogether and was replaced by an almost white translucency. A chemical explanation of the blue-green colour of Swansea duck-egg porcelain has been advanced recently by Edwards (*Swansea and Nantgarw Porcelains: A Scientific Reappraisal*, 2017a) based upon the electronic charge transfer absorptions which occur between the heavy transition metal impurities present in associated minerals in the china clay and bone ash components in the porcelain paste which absorb specific wavelengths of incident light in the visible region of the electromagnetic spectrum. Hence, the green translucencies observed by transmitted light in Figs. 1.3 and 1.10 would be expected to arise from charge-transfer absorptions of red and blue/violet wavelengths in the visible spectrum: the wavelength maximum (green colour) and intensity of the observed transmitted radiation (giving rise to the depth of colour observed) would be dependent upon the number of sites and concentration of heavy metal impurities in the fired specimen.

1.4 The Rise of Chemical Analysis of Welsh Porcelains

Several analyses carried out for Swansea and Nantgarw porcelains have been reported in literature and these have been frequently cited in later texts on Welsh porcelains; these are itemised cumulatively and chronologically as follows and require further description to facilitate the effective comparison of information datasets between the analyses reported or discussed and the type of specimens that have been used:

1. Sir Arthur Church, 1894; recounting his *Cantor Lectures*, 1881.
2. Herbert Eccles, 1914.
3. Herbert Eccles and Bernard Rackham, 1922.

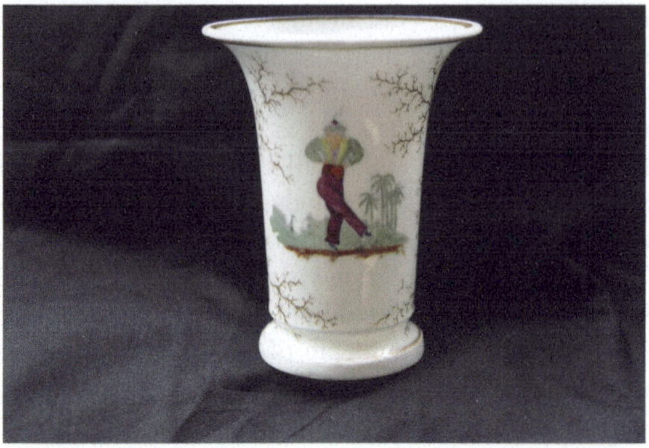

Fig. 1.9 Swansea porcelain spill vase, duck-egg translucency, ca. 1817–1820; chinoiserie and seaweed tendrils decoration attributed to characteristic local decoration by William Billingsley. Private collection

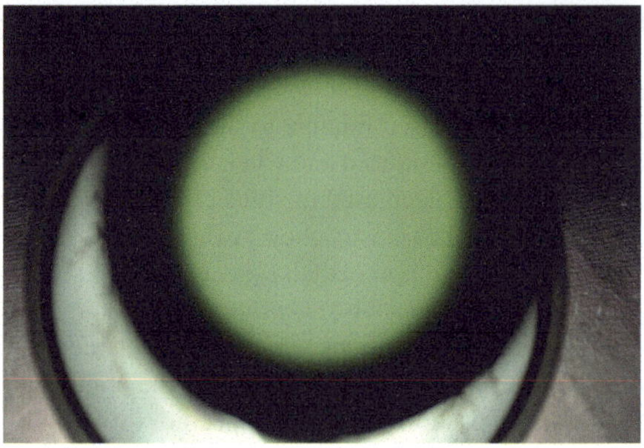

Fig. 1.10 Swansea porcelain spill vase, shown in Fig. 1.7, photographed in transmitted light showing a deep, blue-green duck-egg colour. Private collection

4. Ernest Morton Nance, 1942.
5. Dr. William John, 1948, 1958.
6. "Jimmy" Jones and Sir Leslie Joseph, 1988.
7. Professor Michael Tite and M. Bimson, 1991.
8. Professor Victor Owen, John Wilstead, Rheinallt Williams and Terence Day, 1998.
9. Professor Victor Owen and Michelle Morrison, 1999.
10. Dr. Michael Hillis, 2005.
11. Professor Howell Edwards, 2015.
12. Professor Howell Edwards, 2017.

Thus far, only Eccles (2 and 3), Tite (7) and Owen (8 and 9) have actually reported their detailed analyses of Swansea and/or Nantgarw porcelains: Church (1) does not provide any analytical data at all in support of his analysis of one Swansea and one Nantgarw piece which he has claimed and described in his book (*English Porcelain*, 1894) and, indeed, it appears that he has actually only reported the analyses of the porcelains of Chelsea and Bow and has possibly cited only the available compositional recipes for the other factories, as no precise analytical details are provided (Pollard, *Letters from China: A History of the Origins of Chemical Analysis of Ceramics*, 2015). Herbert Eccles in 1914 analysed just one Swansea piece (recounted in the *Catalogue of the Glynn Vivian Loan Exhibition,* 1914: a centenary exhibition commemorating the establishment of porcelain manufacture at Swansea in 1814) and followed this with the analysis of two Nantgarw and three Swansea pieces in 1922 (Eccles and Rackham, *Analysed Specimens of English Porcelain*, 1922) from the Victoria & Albert Museum in South Kensington, London. Tite and Bimson analysed three Nantgarw pieces in 1991, Owen et al. reported analyses in 1998 on three Nantgarw and three Swansea pieces and Owen and Morrison (1999) analysed a further ten Nantgarw pieces, one potential sherd of Nantgarw glass and one perfect finished and decorated Nantgarw plate from the Nantgarw China Works Museum. The Eccles (1914) analysis was performed on a Swansea plate belonging to the *Gibbins* service from the *Exhibition of Loan China at the Glynn Vivian Art Gallery in Swansea in June 1914*, and cited in Jenkins (*Swansea Porcelain*, 1970) and in Nance (*The Pottery and Porcelain of Swansea and Nantgarw*, 1942, pp. 466–468), which was staged to celebrate the centenary of the first manufacture of porcelain in Swansea by Lewis Dillwyn in 1814, for which some 500 pieces of Swansea pottery and porcelain were assembled together for the very first time since the factory ceased production. This solitary analysis of Herbert Eccles is rather important historically in that it is actually the very first report of recorded chemical analytical data obtained from Swansea porcelain in the literature, but despite this, it seems to have escaped attention and citation in many follow-up articles. It should be noted that although Sir Arthur Church is credited universally with reporting the first analysis of Welsh porcelain, he actually did not actually release any data for his specimens in his book, unlike his detailed work on Chelsea and Bow porcelains in the same publication (Church, *English Porcelain*, 1894)—which themselves suffer from a paucity of information about how the analyses were actually performed! In this early, seminal work Sir Arthur Church describes the history of the manufacturing operations and his observations on porcelain collections in the South Kensington Museum (later The Victoria & Albert Museum) and he comprehensively surveys the porcelains from 23 British factories operating in the 18th and 19th Centuries, comprising Chelsea, Bow, Derby, Worcester, Plymouth, Bristol, Liverpool, Longton Hall, New Hall, Davenport, Minton, Spode, Wedgwood, Lowestoft, Brancas-Lauraguais, Caughley, Coalport, Pinxton, Church Gresley, Rockingham, Nantgarw and Swansea. Of these, Church cites specific analytical chemical data for several nominated specimens of Chelsea and Bow, mentioning briefly Bristol and Brancas-Lauraguais, and giving the associated recipe formulations for Liverpool, Coalport (glaze only) and again for Bow. A possible conclusion, which has not been advanced hitherto, is that Church obtained some of his compositional data from recipes published in the appropriate factory records and which were freely available in contemporary museum archival documentation: the absence of any information from Nantgarw and Swansea would then be satisfactorily explained by the fact that Nantgarw had not published any such information hitherto and Dillwyn's work books (Appendix B), which provided this information for Swansea, were in private ownership and not lodged with the Victoria and Albert Museum (South Kensington) until the 1920s, significantly later than the publication date of Church's work in 1894. It has been already been noted earlier that in 1885, some nine years before Church published his book, Professor Mellor did use a formulation provided by Taylor for the manufacture of "Nantgarw" porcelain—which to all accounts was identical visually and texturally with the original—so, clearly, this recipe was available and in but a limited circulation at that time. Rarity of the specimens in the Museum collections cannot be cited to provide a reason for Church not analyzing the Swansea and Nantgarw specimens, for example, since he reports his analyses of rare Chelsea figures and a solitary surviving but damaged piece of Brancas-Lauraguais porcelain from the Alexandra Palace Fire in 1873 (Soden-Smith, *Catalogue of the Loan Exhibition of Pottery and Porcelain at Alexandra Palace*, 1873). Despite this assertion, Ramsay and Ramsay (2002) cite some lecture presentations (Church, *Cantor Lectures*, 1881) from Church during the early- to mid-1880s in which it appears that he does provide verbal analytical data for some factories during his lectures, but in common with many such investigations earlier in the 19th Century these were not published. At the birth of chemical analysis of archaeological artefacts in the late 18th Century, the credit for the some of the first analyses is occasionally given to the French chemist Vauquelin in the 1790s; these were not published, so it seems that Sir Humphry Davy's published presentation to the Royal Society of London in 1815 (Davy, *Philosophical Transactions of the Royal Society*, 1815) on his chemical analyses of fresco fragments from Pompeii must take precedence in this respect.

The first recorded analysis of Welsh porcelains is also especially interesting for Swansea porcelain analysis because the Swansea specimen analysed by Eccles in 1914 (*GVLE*, 1914) was an example from the so-called *Gibbins* service, with a blue-and-white transfer patterned decoration, having a duck-egg porcelain body impressed with the mark BEVINGTON & CO and having also a red stencilled SWANSEA mark. The china is believed to not have the usual translucency expected for the finest duck-egg body which established the reputation of Lewis Dillwyn (Edwards, *Swansea Porcelain: The Translucent Duck-Egg Vision of Lewis Dillwyn*, 2017c) and is rather coarse in texture by comparison (being described by observers as "rather gritty"), which has been attributed to the inefficient grinding of the materials before firing. Evidently, this documentary piece provides a manifestly unique example of potentially true Bevington porcelain—indicating that they may well have manufactured porcelain at Swansea, albeit perhaps in very minimal quantities and possibly as a trial effort only, and did not just rely upon decorating stock remaining from Dillwyn's proprietorship, as has been supposed by several authorities. An alternative, more likely and tenable explanation is that the Bevingtons acquired some limited production and empirical trial porcelain examples from Dillwyn's earlier experiments and managed to create a service from these for sale, However, what this hypothesis means forensically and analytically is that the analytical data reported by Herbert Eccles on this specimen cannot therefore be truly regarded as absolutely typical of the best Swansea duck-egg porcelain paste as perfected by Lewis Dillwyn at the height of the Swansea factory production. A comparison of the analytical data between this earlier analysis (GVLE, 1914) and that of the best Swansea duck-egg (SW23) porcelain produced between 1816 and 1819 will reveal the chemical nature of this discrepancy and support this hypothesis—and this aspect is discussed in detail later. Apart from this solitary example and the later work of Eccles and Rackham in 1922, the only two examples recorded of finished Swansea or Nantgarw porcelains being studied are the single plate from the Swansea *Biddulph* service form the Royal Institution, Swansea, analysed by Owen et al. in 1998 and one piece of decorated Nantgarw china from the Nantgarw China Works Museum studied by Owen and Morrison in 1999. Therefore, the sum total of finished Swansea and Nantgarw porcelains analysed to date comprises Eccles (1 Swansea), Eccles and Rackham (2 Nantgarw and 3 Swansea), Owen et al. (1 Swansea) and Owen and Morrison (1 Nantgarw)—in all some eight pieces, several of which were already damaged or extensively broken prior to the analyses being undertaken.

Sir Arthur Church's book (*English Porcelain*, 1894) does include one example each of Swansea and Nantgarw porcelain illustrated in his Figs. 6.29 and 6.30 lodged in that text, these being a saucer and a cabinet cup from the Schreiber Collection in the Victoria and Albert Museum (South Kensington), which had been donated to the Museum by Lady Charlotte Schreiber in 1868 and 1878, respectively. A description accompanies the items on pages 93–96 of his book (Church, *English Porcelain*, 1894), which outlines the creation of the manufactories and their subsequent downfall historically. No analytical data is given for either Swansea or Nantgarw pieces, although mention is made earlier in the text of Chelsea and Bow "analytical" compositions but without the provision of any accompanying details of the methodology and experimentation used to derive these data. Church's book, written so chronologically close to the heyday or recent closure of many of these porcelain factories is a wealth of background information, which, for example, correctly points out that the Chelsea, Bow, Liverpool and Caughley factories all recorded the use of calcined bone ash in their porcelain formulations from the 1760s onwards and he comments correctly that "*the frequent attribution to Josiah Spode of his first introduction of bone ash into the paste of English porcelains in 1797–1800 must be regarded as destitute of any basis of fact*". Unfortunately, the accounts given by Church of the Swansea and Nantgarw factories in comparison are patently incorrect in several important details. For example, Church states that after the closure of the first phase start-up of Nantgarw in 1814, the kilns, equipment and stock were moved to Swansea and after the eventual closure of the Swansea and Nantgarw factories in 1820, John Rose of Coalport purchased the appliances, moulds and equipment and transferred these to Coalport, employing William Billingsley and Samuel Walker in this process. As pointed out earlier, a partial credence can be given to this allegation in that some years later the Coalport China Works adopted a new rebus for marking its products: namely, an ampersand (Fig. 1.8) enclosing the letters C, S and N, which have been translated as Caughley, Swansea and Nantgarw—emphasising the recent acquisitions of these three china works by John Rose and their subsequent incorporation into John Rose's Coalport factory. The letter S has alternatively been taken to refer to "Salopian", but this seems possibly tautological as the Caughley factory is also known as Salopian? Church makes no mention of the start-up at Nantgarw of the so-called second phase in 1817, which presumably according to his hypothesis would have necessitated the re-location of the kilns and equipment again to Nantgarw from Swansea and a consequent and resultant hiatus in production of porcelain at the Swansea site! He also maintains that china continued to be made in Nantgarw after the departure of Billingsley and Walker to Coalport in 1819, but he admits that no examples of this porcelain have yet been identified up to that time; clearly, here he is confusing the documented decoration of the residual porcelain stock at Nantgarw by Thomas Pardoe for sale in the period 1820–1823 with an assumed re-opening of the manufactory production to prepare new porcelain offered for sale at the auctions. There is also a possibility of historical confusion in this surmise because it is on record that a William Pardoe reopened the factory premises at Nantgarw in the late 1830s for several years, but for the express purpose of manufacturing there earthenware clay pipes and not porcelain. Church also states that, as a consequence of the transfer of Billingsley and Walker from Nantgarw to Swansea, the Nantgarw recipe was adopted at Swansea and, thus, that the Swansea porcelain paste is analytically identical to that of Nantgarw porcelain. This assertion, of course, can now be checked analytically in the experiments that will be described later There was a rumour in circulation that Samuel Walker did consider restarting the Nantgarw kilns immediately after the death of William Billingsley in 1828, but without financial backing he realised this would not be a viable proposition commercially: this means of course that in 1828, the Nantgarw hardware (moulds, kilns, and grinders) must still have been extant in Nantgarw at that time and had not been removed by John Rose to Coalport as was claimed allegedly some years earlier!

The main visual difference between Swansea and Nantgarw porcelains in the opinion of Sir Arthur Church was the highly embossed nature of the Swansea porcelain articles, which he maintains were all decorated locally by William Weston Young, and that Swansea porcelain was manufactured only between 1814 and 1817, after which Dillwyn signed over the factory to the Bevingtons. These remarks are rather fanciful and can now be dismissed factually, along with their totally inaccurate chronology, by cross-referencing with existing documentation: the true situation is much more complex in context and the interested reader is referred to the holistic research carried out by Edwards (*Swansea and Nantgarw Porcelains: A Scientific Reappraisal*, 2017a), which corrects many of these statements and others made in the earlier literature from which several incorrect assertions have followed. A particularly comprehensive study of the border moulding designs seen on Nantgarw and Swansea china has been undertaken by Jones and Joseph (*Swansea Porcelain*, 1988) and from their analysis several discernible features in the moulding can be identified to discriminate between the two factories: the conclusion is that the Swansea and Nantgarw border mouldings are not identical and therefore any suggestion that the moulds from one factory were used at the other are not tenable. However, Church does make an interesting statement, in passing, about the total loss of 24 Nantgarw and 27 Swansea porcelain pieces of exceptional quality that were on exhibition in the Alexandra Palace (*The People's Palace*), which suffered a devastating fire in 1873, just 16 days after it had opened to the public for their recreation and education: all together, some 4700 pieces of important historic British china were totally destroyed in this fire (Soden-Smith, *Catalogue of the Ceramics at the Alexandra Palace Exhibition*, 1873). The loss through fire of collections of porcelain seems sadly to have been not unusual in the 19th century, as a later fire in 1893 at the family home of the Vivians in Swansea, who were major supporters and clients of the Swansea china manufactory some eighty years earlier, destroyed it is estimated hundreds of many of the finest quality Swansea porcelain pieces ever made, including several important exquisitely decorated and complete dinner and dessert services, of which only single examples now remain as these escaped destruction fortuitously through being away on loan as exemplars at an exhibition in Cardiff at that time (Edwards, *Swansea and Nantgarw Porcelains: A Scientific Reappraisal*, 2017a; Nance, *The Pottery and Porcelain of Swansea and Nantgarw*, 1942).

A more detailed comparison of the experimental analytical data and the conclusions derived from them needs now to be undertaken in an attempt at understanding the superior translucency of the soft paste porcelain manufactured at Swansea and Nantgarw, which was never matched by its esteemed contemporary competitors, such as Barr, Flight & Barr at Worcester and Robert Bloor at Derby: an important part of this analytical exercise will be a consideration of the solid state chemical reactions which have taken place during the firing of the empirical body compositions employed at the high kiln operational temperatures at Swansea and Nantgarw, which create molecular and molecular-ionic materials affecting the stability and translucency of the resultant china.

1.5 Location of Nantgarw and Swansea Sites and Transportation Issues

The situation of the Nantgarw site and its access to a first class communications system in the early 19th Century has already been mentioned: the importance of a communications infrastructure for the viability of a commercial ceramics production enterprise has been rather ignored until now for the importation of raw materials and the export of the finished porcelain articles, so it is opportune to consider this here and the influence that pertained upon the manufacture, production and distribution of quality porcelain. Josiah Wedgwood bemoaned the fact that between 75 and 80% of the final cost of his finished ceramics despatched for sale in London from Etruria in the Staffordshire Potteries could be attributed to transportation charges (Thomas, *The Rise of the Staffordshire Potteries*, 1971, p. 85). Most transportation involved the use of carriages on the road system, for which packages sent by mail coaches under armed guard were the safest form of express transportation but also the most expensive: limited to an individual weight of one pound (approximately 0.45 k), porcelain articles were occasionally despatched using this means. Joseph Lygo, the Derby China Works agent in London, has documented that a post-haste journey by mail coach from Derby to London took some 20 h at an average speed of 5 mph, whereas the alternative heavy haulage "waggons" were cheaper but suffered from the frequent loss of the articles consigned to them—and from which consignments often took up to four weeks to arrive anyway! Hence, for the delivery of the prestigious

Duke of Northumberland service to his warehouse in London in 1790, Lygo specified to the Derby China Works that the mail coaches must be used for transport to London from Derby and that William Duesbury should bear the additional cost involved. The situation seems not to have improved significantly into the early 19th Century since a very prestigious Royal service ordered by King William IV in 1835 (Cox & Cox , 2001) from the Rockingham factory in Swinton, Yorkshire, was collected by a detachment of the King's Troop, Royal Horse Artillery, who led a train of pack horses, each fitted with velvet lined mahogany panniers in which were secured individual examples of this fine porcelain service for carriage to London. This service never arrived during King William IV's lifetime and was eventually used at the Coronation Banquet of Queen Victoria in Windsor Castle in 1837, where it still resides. Examples of pattern plates from this exquisite Royal service are now eagerly sought by museums and collectors of Rockingham porcelain and several are illustrated in Cox and Cox (*Rockingham Porcelain 1745–1842*, 2001); sadly, it seems that the Rockingham factory never received payment for this prestigious service, which bankrupted them financially and resulted in the enforced factory closure soon afterwards.

It should also be appreciated that the usual carriage of porcelain items on the so-called "waggons", although relatively much cheaper than the mail coach option, also resulted in a demonstrably more hazardous and rougher journey even though the tolls had still to be paid for use of the turnpikes. William Billingsley reckoned that some 15% of his finished porcelain at Derby and Pinxton never survived the delivery trips to his clients by this mode of transport and it is thought that this could be an underestimate in several cases. Joseph Lygo, the London agent for the Derby porcelain manufactory, comments that he requested that the Derby China Works supplied him with an excess in number of decorated pieces for service commissions to allow for some breakage in transit and this would facilitate his final making-up of a complete service in London. A documented example of this was evident for a dessert service commissioned in 1790 by Thomas Johnes Esq. of Hafod, a very influential patron of Derby porcelain, who ordered through Lygo in London a special service of Derby porcelain and for which some two hundred items were requested to be sent from Derby to London by Lygo. Lygo then took it upon himself to assemble the requisite service composition for Johnes, noting along the way that Derby had included several inferior pieces in their batch painted by second rate artists to complete the extended commission by the required dateline. He stated that he only selected the very best examples for Mr. Johnes' particular service, incorporating landscapes painted by Zacariah Boreman, flowers by William Billingsley, figures by Josh Banford and fruit by George Complin, accompanied by superior gilding from Messrs. Soar, Yates and Cooper (Anderson, *Derby Porcelain*, 2000). Perhaps this practice initiated the request from discerning clients that artistry on the commissioned services be executed by specifically named and esteemed artists, such as Zacariah Boreman on the *Hafod* landscape scenes service, Edward Withers on the *Duke of Northumberland* floral service and William Billingsley on the *Prince of Wales* and *Earl Camden* floral services—all of which are considered to be *crème de la crème* examples of 18th Century-decorated Derby porcelain.

The transportation problem outlined above also extended to the importation of essential raw materials for the factories, which will be of particular relevance for the analytical discussions to follow as the sourcing of raw materials was no easy matter and the influence of mineral and other impurities in the raw materials was critical for porcelain translucency. Early experimental notes that survive from factories such as Derby indicate that the owners scoured the country, and even the wider Europe, for the best raw materials. An example is provided by charcoal for effecting enhanced and superior temperatures in the firing and glost kilns—one needs to exercise care in assessing early letters and documentation relating to the acquisition and transport of charcoal since there is much confusion in the original records by purchasers of coal and charcoal for kiln firing as these were often referred to simply and rather loosely as "*coal*" or even a hybridised "*CharCoal*". A note in William Duesbury's Derby China Works workbook refers to the selective purchase of several cords of charcoal from Meissen in Saxony as an alternative to his usual supplier in Newcastle, and calcined bones from London, when more local variants were clearly available presumably more cheaply (Anderson, *Derby Porcelain*, 2000). In the latter case, there was much legendary folklore in that ox bones were the best for calcination, followed by cow bones but horse bones should never be used—and even calcined fish bones were deemed to give better quality bone ash than horse bones! For the importation of raw materials, waterborne transport was significantly cheaper than road carriage, mainly because of the heavy tolls imposed on the turnpikes and minor roads, despite the longer delivery times involved: here, the Midlands pottery factories suffered in that the River Severn was the only navigable English river that could be negotiated without floodgates, locks and weirs, and the canal system was not as extensive as it transpired later. The opening of a new Derby porcelain outlet warehouse in Bath to cater for the Georgian society and aristocracy who "seasoned" there was encouraged by Joseph Lygo to increase their competitive sales: what was not immediately realised, however, was the influence that increased transportation costs and the poorer infrastructure around Bath would have on the decorated Derby porcelain exemplars being exhibited there. Hence, greater losses in breakage of quality porcelain through bad carriage was incurred and the increased costs of road carriage, even when an alternative initial sea route from London to Weymouth was tried, resulted in this mission being pronounced a failure eventually. An attempt to transfer artists from the Derby factory to Bath or London at peak times for the decoration of unfinished porcelain shipped there in the white was also not economically viable because a one-way trip by mail coach cost approximately one week's wages for a skilled artisan and the alternative "waggon" trip was too arduous and debilitating. Hence, the renowned figure modeller Jean-Jacques Spangler, returning to Derby in 1795 from a London warehouse visit, expressed a preference for walking the route over his alternative use of a ponderous and rough carriage ride, which William Duesbury agreed was also much more cost-effective (Anderson, *Derby Porcelain*, 2000)! Hence, too, it explains why William Billingsley, who made his journey from Worcester to Nantgarw to set up the factory there in 1811, undertook part of the trip by waterborne transport to Bristol and thence on foot to Nantgarw, and why his artist friend William Pegg and his wife joined the

Billingsleys at Nantgarw after walking all the way from Derby, an arduous trek which William Pegg describes evocatively in his contemporary correspondence.

It can now be appreciated that William Billingsley's start-up venture in Nantgarw was especially well selected because of the excellent transportation infrastructure already in place there: the Glamorgan Canal had just been completed, situated adjacent to the porcelain factory site and linking Cardiff docks some seven miles away to the Rhondda Valleys, and capable of taking heavy-tonnage, horse-drawn barges. The Welsh coalfield and the Rhondda Valleys gave a local and readily accessible supply of superb high quality steam coal, with a the highest ratio of output thermal heat upon combustion versus weight, and this source remained the preferred material later in the Century for firing the boilers of the ironclads and dreadnoughts of the Royal Navy and illustrious passenger liners well into the 1920s. Local ox bones were calcined and ground by a Mr. David Jones in his local mill to the exacting perfection required by Billingsley; this was realised to be a key operation in the manufacture of defect-free translucent porcelain and has been cited by Church (*English Porcelain*, 1894) with particular reference to imperfections in mixing in early Chelsea porcelain giving rise to translucent "moons" when viewed by transmitted light. The removal of organic detritus from the calcined bones was also appreciated by Lewis Dillwyn at Swansea and William Duesbury at Derby, who both made great efforts to secure the best quality material for use as a component in their porcelains. China and ball clays were sourced from Cornwall, fine quartz sand from Lynn in Norfolk and cobalt blue (smalt) from Bristol—all of these could be imported by sea through Cardiff. Finally, the finished articles could be exported via Cardiff down the Bristol Channel and around the Lizard to London directly—the finished porcelain, thus, would have no potentially expensive weight restrictions and hazardous transit resulting in further losses for road carriage and could be crated up effectively and loaded at the Nantgarw manufactory site directly onto the canal barges. Similarly, Lewis Dillwyn at Swansea had the benefit of the deep-water Swansea docks for his input and export of materials and finished porcelain, just a short distance from the factory site. It has been stated (Anderson, *Derby Porcelain*, 2000) that Swansea would not have been subject to the breakages incurred in transportation of their finished porcelains in 1821–1822 since most of the porcelain was decorated locally anyway. This is true, but it should also be remembered that Swansea porcelain, unlike Nantgarw, had five major retail outlets in London, including the renowned Apsley Pellatt and Green of St Paul's, and they would therefore have had to send significant amounts of finished duck-egg china to supply their London retailers up to 1819, when porcelain production ceased and Dillwyn sold out to the Bevingtons. What Anderson seems to be referring to is the regime imposed by the Bevingtons to decorate the residual stock left in the white at Swansea for sale locally and at the subsequent auctions after 1823: before that, however, there is much documentation to support a vigorous and active selling outlet of Swansea porcelains in London and indeed it would be incorrect to assume that most Swansea porcelain was locally decorated and sold there—after all, Dillwyn's disastrous attempt to replace his exquisite duck-egg translucent porcelain with the muddier but more robust trident body was not appreciated by his London retailers at all and, as a result, the downfall of his Swansea enterprise was assured, stressing the importance of the London outlet for Swansea china. In this, it can be argued that Swansea and Nantgarw were probably similar to Derby porcelains in that some 85–90% of the factory output was sold through a London agent or retailer (Godden, *Godden's Guide to European Porcelain*, 1993, p. 21), even though there may be some discrepancy between the amounts that would have been decorated locally and sent to London for subsequent sale—here, much of the Nantgarw output was supplied solely to John Mortlock in London in the white for decoration in the ateliers of Robins and Randall or John Sims, whilst much Swansea and Derby in contrast was still decorated locally and then subsequently sold in London.

References

J.A. Anderson, *Derby Porcelain and the Early English Fine Ceramic Industry*, Ph.D. Thesis, University of Leicester, UK, Oct 2000

E.A. Barker, *Pottery and Porcelain in the United States* (New York, 1893), p. 178

W. Bemrose, *Bow, Chelsea and Derby Porcelains* (Bemrose, London, 1898)

W. Burton, *Porcelain, A Sketch of its Nature, Art and Manufacture* (B.T. Batsford, London, 1906)

W.B. Chaffers, *Marks and Monograms on Pottery and Porcelain with Historical Notes on Each Manufactory* (J. Davy & Sons, London, 1863) (Kessinger Legacy Reprints, Kessinger Publishing, Whitefish Montana USA, 2010)

A.H. Church, Cantor Lectures on Some Points of Contact Between the Scientific and Artistic Aspects of Pottery and Porcelain, Lecture IV. *Journal of the Society of the Arts* (1880), pp. 126–129. (Extended in monograph by Trounce Publishers, London, 1881)

Sir A.H. Church, in *English Porcelain: A Handbook to the China Made in England During the 18th Century as Illustrated by Specimens Chiefly in the National Collection.* A South Kensington Museum Handbook (Chapman & Hall Ltd., London, 1894)

P. Colomban, The Destructive/Non-destructive Identification of Enameled Pottery, Glass Artifacts and Associated Pigments—A Brief Overview. Arts **2**, 77–110 (2013)

P. Colomban, V. Milande, H. Lucas, On-site Raman Analysis of Medici Porcelains. J. Raman Spectrosc. **35**, 68–72 (2004)

A. Cox, A. Cox, *Rockingham Porcelain* (Antique Collectors Club Publishing, Woodbridge, Suffolk, 2001)

Sir H. Davy, Some Experiments and Observations on the Colours Used in Paintings by the Ancients. Philos. Trans. R. Soc. **105**, 97–124 (1815)

Sir H.T. de la Beche, T. Reeks, F.W. Rudler, *Catalogue of Specimens in the Museum of Practical Geology, Jermyn Street, London, Illustrative of the Composition and Manufacture of British Pottery and Porcelain from the Occupation of Britain by the Romans to the Present Time*, 3rd edn., ed. by G. Eyre, W. Spottiswoode (For the HMSO, London, 1876)

L.W. Dillwyn, *Notes on the Experimental Production of Swansea Porcelain Bodies and Glazes.* Made by Lewis Weston Dillwyn with Samuel Walker at the Swansea China Works Between 1815 and 1817. Presented to the Library of the Victoria &Albert Museum, South Kensington, London by John Campbell in 1920 (Reproduced in Eccles and Rackham,*Analysed Specimens of English Porcelain*, 1922, see reference)

R. Dossie, *The Handmaid to The Arts*, vol II (Published by J. Nourse at The Lamb, opposite Katherine Street, The Strand, London, 1758), p. 342

H. Eccles, B. Rackham, *Analysed Specimens of English Porcelain in the V&A Museum Collection* (Victoria & Albert Museum, South Kensington, London, 1922)

H.G.M. Edwards, Swansea and Nantgarw Porcelain Bodies Based on Analytical Evidence: A Case Study, in *Encyclopaedia of Analytical Chemistry*, ed. by R. Meyers, Y. Ozaki (Chichester, UK, Wiley, 2015a)
H.G.M. Edwards, Historical Pigments, in *Encyclopaedia of Analytical Chemistry*, ed. by R. Meyers, Y. Ozaki (Chichester, UK, Wiley, 2015b)
H.G.M. Edwards, *Swansea and Nantgarw Porcelain: A Scientific Reappraisal* (Springer Publishing, Dordrecht, The Netherlands, 2017a)
H.G.M. Edwards, *Nantgarw Porcelain: The Pursuit of Perfection* (Penrose Antiques Ltd). Short Guides, Series ed. by M.D. Denyer (Penrose Antiques Ltd., Thornton, West Yorkshire, UK, 2017b). ISBN: 978-0-244-90654-2
H.G.M. Edwards, *Swansea Porcelain: The Duck-Egg Translucent Vision of Lewis Dillwyn* (Penrose Antiques Ltd., Short Guides, Thornton, West Yorkshire, UK, 2017c). ISBN: 9780244325787
H.G.M. Edwards, *Derby Porcelain: The Golden Years, 1780–1830* (Penrose Antiques Ltd., Short Guides, Thornton, West Yorkshire, UK, 2018)
H.G.M. Edwards, M.C.T. Denyer, *William Billingsley The Enigmatic Porcelain Artist, Decorator and Manufacturer* (Penrose Antiques Ltd. Short Guides, Neopubli, Berlin, 2016), pp. 656–661, 2004, ISBN: 978-3-7418-6802-3
C.L. Exley, *The Pinxton China Factory* (Mr. & Mrs. R Coke-Steele Publishers, Derby, 1963)
I. Freestone, in *Science and Early British Porcelain*. Proceedings of the Sixth Conference and Exhibition of the European Ceramics Society, London, June 1999, International Ceramics Societies, vol 1 (published in 2000), pp. 11–17
G.A. Godden, *Godden's Guide to European Porcelain* (Barrie & Jenkins, London, 1993), p. 21
J. Haslem, *The Old Derby China Factory* (George Bell, London, 1876)
J. Haslem, *A Catalogue of China* (R. Keene, Derby, 1879)
M. Hillis, The Development of Welsh Porcelain Bodies, in *Welsh Ceramics in Context, Part II*, ed. by J. Gray (Swansea, Royal Institution of South Wales, 2005), pp. 170–192
E. Jenkins, *Swansea Porcelain* (D. Brown Publishers, Cowbridge, UK, 1970)
W.D. John, *Nantgarw Porcelain* (Ceramic Book Co., Newport, 1948)
W.D. John, *Swansea Porcelain* (Ceramic Book Co., Newport, 1958)
W.D. John, *William Billingsley* (Ceramic Book Co., Newport, 1968)
W.D. John, G.J. Coombes, K. Coombes, *The Nantgarw Porcelain Album* (Ceramic Book Co., Newport, 1975)
A.E. Jones, Sir L. Joseph, *Swansea Porcelain: Shapes and Decoration* (D. Brown and Sons Ltd, Cowbridge, 1988)
C. Jorg, P. Wilson, *Collection of the Rijksmuseum, Amsterdam: The Ming and Qing Dynasties* (Rijksmuseum Amsterdam Press, Amsterdam, 1997)
I.B. McAbe, *Orientalism in Early Modern France* (Berg Publishing, Oxford, 2008)
E.M. Nance, *The Pottery and Porcelain of Swansea and Nantgarw* (B.T. Batsford Ltd., London, 1942)
W. Nicholson, *A Dictionary of Practical and Theoretical Chemistry, with its Application to the Arts and Manufactures and to the Explanation of the Phenomena of Nature: Including the Latest Discoveries of the Precise State of Knowledge in these Subjects* (Richard Phillips, London, 1808)
J.V. Owen, A New Classification Scheme for 18th Century American and English Soft Paste Porcelain, in *Ceramics in America*, ed. by R. Hunter (USA, Chipstone Foundation, Milwaukee, Wisconsin, 2002), pp. 45–61
J.V. Owen, M.L. Morrison, Sagged Phosphatic Nantgarw Porcelain (ca. 1813–1820): Casualty of Overfiring or a Fertile Paste?. Geoarchaeology **14**, 313–332 (1999)
J.V. Owen, J.O. Wilstead, R.W. Williams, T.E. Day, A Tale of Two Cities: Compositional Characteristics of Some Nantgarw and Swansea Porcelains and Their Implications for Kiln Wastage. J. Archaeol. Sci. **25**, 359–375 (1998)
A.M. Pollard, Letters from China: A History of the Origins of the Chemical Analysis of Ceramics. AMBIX (Society for the History of Archaeology and Chemistry) **62**, 50–71 (2015)
M. Polo. *Il Milione,* cap. CLVIII, dell edizione a cuva di L.F. Benedetto, transl. 1928

E.G. Ramsay, W.R.H. Ramsay, Bow First Patent Porcelain: New Discoveries in Science and Art. *The Antiques Magazine* (Brant Publications, New York, September issue, 2006), pp. 122–127

E.G. Ramsay, W.R.H Ramsay, in *Bow: Britain's Pioneering Porcelain Manufactory of the 18th Century.* The International Ceramics Fair & Seminar (Park Lane Hotel, London, 2007a), pp. 1–16

W.R.H. Ramsay, E.G. Ramsay, A Classification of Bow Porcelain from First Patent to Closure: c.1743–1774. Proc. R. Soc. Vic. **119**(1), 1–68 (2007b). ISSN 0035-9211-1-168

W.R.H. Ramsay, E.G. Ramsay, *The Evolution and Compositional Development of English Porcelains from the 16th C to Lund's Bristol c. 1750 and Worcester c. 1752—the Golden Chain* (Invercargill, New Zealand, 2017)

W.R.H. Ramsay, A. Gabszewicz, E.G. Ramsay, The Chemistry of A-marked Porcelain and its Relation to the Edward Heylyn and Thomas Frye Patent of 1744. Trans. Engl. Ceram. Circle **18**, 264–283 (2001)

W.R.H. Ramsay, G. Hill, E.G. Ramsay, Recreation of the 1744 Heylyn and Frye Ceramic Patent Wares using Cherokee Clay: Implications for Raw Materials, Kiln Conditions and the Earliest English Porcelain Production. Geoarchaeology **19**, 635–655 (2004)

W.R.H. Ramsay, F.A. Davenport, E.G. Ramsay, The 1744 Ceramic patent of Heylyn and Frye: Unworkable *Unaker* Formula or Landmark Document in the History of English Ceramics. Proc. R. Soc. Vic. **118**, 11–34 (2006)

S. Shaw, *The Chemistry of the Several Natural and Artificial Heterogeneous Compounds Used in Manufacturing Porcelain, Glass and Pottery* (Scott, Greenwood & Son, London, 1837). Re-issued in its original form in 1900, p. 713, 1900

R.K. Soden-Smith, *Catalogue of the Collection of English Pottery and Porcelain Exhibition On Loan at the Alexandra Palace* (R.K. Burt and Co., London, 1873)

J. Thomas, *The Rise of the Staffordshire Potteries (The Origins of Industry)*, Adams & Dart / Jupiter Books, London, p.85, 1971

J. Taylor, *The Complete Practical Potter* (Shelton, Stoke-upon-Trent, 1847)

M.S. Tite, M. Bimson, A Technological Study of English Porcelains. Archaeometry **33**, 3–27 (1991)

W. Turner, *The Ceramics of Swansea and Nantgarw* (Bemrose & Sons, Old Bailey, London, 1897)

J. Twitchett, *Derby Porcelain* (Barrie and Jenkins, London, 1980)

J. Twitchett, *Derby Porcelain 1748–1848: An Illustrated Guide* (Antique Collectors Club, Woodbridge, Suffolk, 2002)

Chapter 2
Porcelain in the Eighteenth Century and Its Standing in Georgian and Regency Society

Abstract The acquisition of porcelain and the perceived reflection of the standing of the family in Georgian and Regency Society in the 18th and 19th Centuries contributed significantly to the growth of the manufactories. The reverence with which it was treated and responsibility for its care in the household can be correlated with the commissioning of porcelain services and tablewares: here the concept of a "named service" is introduced and the dating of a commission afforded to input directly with any porcelain paste changes. The attribution of a ceramic artist's work to assigned services is strengthened by such information and this can be used to identify the artist's work on other, as yet unrecognised, commissions. The absence of factory pattern books is a major flaw in this application as will be demonstrated with the two factories under consideration here, namely, Swansea and Nantgarw.

Keywords Pattern books · Named services · Desirability of porcelain ownership Commissioned services · Attribution of ceramic artworks · Named artists

It is relevant to consider the role that porcelain had in polite and aristocratic society, which reflects the desirability of ownership of this expensive but rather fragile commodity and also the status of the competitive factories, which supports the striving of each to achieve a market edge in the light of a changing social vogue which imposed some special demands upon the suppliers. For example, the simplicity of the mid- to late- 18th Century Georgian decoration, as exemplified by the *Prince of Wales* service created in Derby in 1786, showing a dessert plate from this same service in Fig. 2.1, gave way to an embellishment of over-decoration accompanying more ostentatious shapes and devices made of porcelain. By the accession of Queen Victoria in 1837, the revived rococo and baroque styles exemplified by the later Bloor Derby, Coalport and Rockingham porcelains reflected a change in taste far removed from the pure forms and decoration of their earlier Georgian and Regency counterparts. An example of the onset of this revived rococo and baroque style is seen in the covered sucrier from a high—quality tea and coffee service made at Coalport in

H. G. M. Edwards, *Nantgarw and Swansea Porcelains*,
https://doi.org/10.1007/978-3-319-77631-6_2

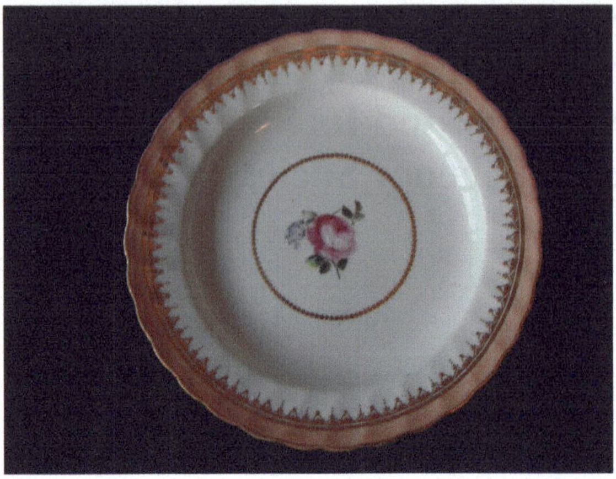

Fig. 2.1 Derby porcelain plate from the Prince of Wales service, 1786, pattern 65, decorated by William Billingsley with a single pink rose and forget-me-nots, dawn pink edging and gilded by William Longden, gilder's number "8" near edge of footrim. Puce mark, cursive D with crown and crossed batons. Private collection

the late 1820s (Fig. 2.2) which is still beautifully decorated and gilded but is much more ostentatious.

It may not be fully appreciated that fine porcelain was treated with deference and respect in society households in Georgian England and that, as a result, the ladies of the household were reluctant to fully delegate the responsibility for its care to their servants. For example, Mrs Susannah Whatman of the Manchester paper-making dynasty, whose husband had the very sizeable income of £6000 per annum in the late 1790s, was directly involved personally in the washing and cleaning of the family porcelain, unlike the family silverware which was the responsibility of the butler (Vickery, *The Gentleman's Daughter*, 2004, p. 149; Whatman, *The Housekeeping Manual of Susannah Whatman*, 1987). The acquisition of vulnerable and friable soft-paste porcelains by clients had resultant and often disruptive social consequences amongst staff and their employers especially when breakages and damage

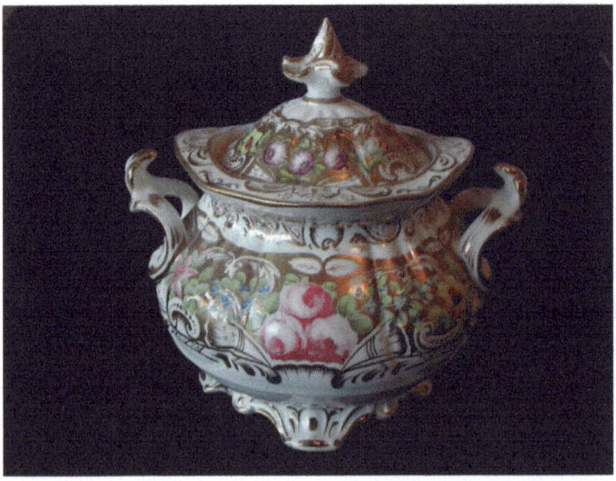

Fig. 2.2 Coalport porcelain sucrier with lid, ca. 1815–1825, sumptuously gilded and decorated in baroque style. Unmarked, but pattern number 2/675 in red enamel written to the base. Service illustrated in Godden, Coalport and Coalbrookdale Porcelains, 1991

occurred: Joseph Lygo has recorded and documented breakages of Derby porcelain among his clients and there exist in the Derby record books details of repairs undertaken to services on behalf of clients in the 1790s: these include Lord Wentworth's icepail, Queen Charlotte's dessert service, the Duchess of Ancaster's service, Lady Grantham's tea service which also required "… *cleaning the discolouration arising from seasoning*". The costs involved in repair, cleaning and replacement with factory items were quite significant—the Prince of Wales had five icepails repaired at 2s each and two new replacement teapots for tea services at 7 guineas per pair invoiced in 1790. The social standing of Derby clients for the provision of these services was listed by Lygo in 1783 according to "*identifiable types*": these included 3 Royalty, 20 Dukes and Lords, 4 Duchesses, 9 Ladies, 8 Knights, 2 Bishops and several Generals, Admirals and MPs. The proximity of Lygo's London warehouse to the Houses of Parliament was especially advantageous for the procurement of porcelain orders and commissions from influential and wealthy clients. This list also gave an indication of the social standing of potential clients who would then normally be visited at home by Joseph Lygo and/or William Duesbury, who sought their support and custom: other clients would be requested to visit the warehouse and place their orders there during the London season, before returning to their country estates. The function of porcelain items as status gifts has also been considered but it is very difficult to assess (Anderson, *Derby Porcelain*, 2000) as the reason for the specific commission is rarely provided in the factory order books. However, a tantalising glimpse of the situation is provided in some correspondence between William Duesbury and Joseph Lygo, which has fortunately been retained in the Derby archives: the Prince of Wales in 1794 presented a dessert set to Queen Charlotte in gratitude for her personal "support" of him in "*recent potentially difficult situations*", Lady Harrington gave some ornamental porcelain to the young Princesses, Mr. Thomas Johnes of Hafod gave a dessert service to the Lord Chancellor and another to Dr. Pittcairne, "*who attends my family but takes no fees*". These gifts were significantly expensive: for example, the latter service mentioned a cost of 40 guineas, approaching the annual salary of a senior member of the clergy at that time.

Like Swansea and Nantgarw, the London outlet for Derby was a critical part of their retail operation—Derby selling some 85–90% of its total factory output there, similar figures certainly applying for Nantgarw and, perhaps, Swansea too. An interesting result of this practice is the ability of an agent to control the composition of a commissioned service, especially where breakages have occurred in transit and the onset of a delivery deadline approached. It has been highlighted (Edwards, *Swansea & Nantgarw Porcelains: A Scientific Reappraisal*, 2017) that too much reliability may have been accorded by researchers hitherto to individual items in a Swansea or Nantgarw service where only several pieces have been marked and even when a reliable and recorded provenance was in place, thereby implying that the whole service then obviously originated from the factory source. For example, it is well known amongst researchers that some pieces from the *Lysaght* Swansea service are believed to be Coalport in origin, particularly dessert dishes and comports of a highly irregular shape for Swansea items. The same situation applies to the *Biddulph* service, the former being decorated en suite in Swansea by Henry Morris and the latter in London by Philip Ballard. An example of a richly decorated and gilded Swansea *Lysaght* service plate is shown in Fig. 2.3. There has also been a suggestion, perhaps originating from disaffected retailing competitors, that John Mortlock substituted some non-Nantgarw items in composite services ordered through his retailing business to make up the specific needs for special client commissions and thereby retailing these associated items at Nantgarw porcelain prices. It is believed that occasionally rare items in dessert or dinner services, such as Nantgarw and Swansea icepails, come into this category, and recent studies of the *Marquess of Exeter* and *Gosforth Castle* Swansea services seem to indicate that the icepails could well be associated items from other factories bought in and decorated en suite to complete the services in the London enamelling ateliers. Of course, as we have related above, breakages did occur in transit and usage and clients would naturally seek replacements for or additions to a favourite and treasured porcelain service at a later date: a classic case is provided by the *Duke of Northumberland* Derby service first ordered in 1785 and decorated with pink roses by Edward Withers, then the finest rose painter working at Derby and the head of their enamelling workshop. In 1790, the Duke of Northumberland decided that he wished to extend his service by the incorporation of soup dishes, so he placed an order with William Duesbury for 24 soup dishes, which were duly decorated en suite with the older service items by William Billingsley who had by then taken over from Edward Withers, his previous instructor at Derby, as the prime rose painter in the Derby China Works enamelling studio at that time. These dishes are now avidly desired as genuine examples of Billingsley's early Derby work by discerning collectors, and they certainly cannot therefore be considered as examples from the hand of Edward Withers because he was recorded as the "named" artist for this service in the factory pattern books. The situation at Swansea and Nantgarw is rather more complex than that provided by this example from Derby as both factories ceased production after only a very few years of engagement in production, so it would not be possible for clients to acquire replacements for broken items or for an additional extended composition for a service from the source factories at a significantly later date. Then, the only solution for such a client would be to add new items from another factory, decorated in a similar style, but perhaps offering an unusual shape or moulding more typical of that factory than would have previously characterised the then defunct two Welsh factories and these may even have acquired thereby a false factory mark! Another example is provided by items from the *Kenyon* Swansea service, which occasionally appear in the auction rooms but with rather different decoration and applied mouldings—these could be considered either generic replacements or attempts to engender fake pieces entering the system but clearly such different pieces cannot be the from the same original service? This account therefore lays the foundation and necessity for an objective chemical analysis programme to be undertaken without prejudice to secure better information about the individual factories and the assessment and attribution of unknown pieces to a particular factory production.

The author finds it strange that such a high-profile ceramics manufacturing operation in the 18th Century never brought with it the support of the Guilds in the City of London—which at the close of the century numbered 72: an early foundation

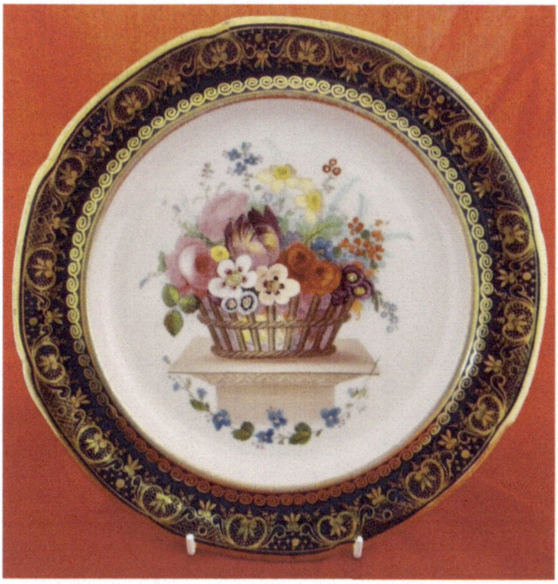

Fig. 2.3 Swansea porcelain dinner plate from the Lysaght service, duck-egg translucency, ca. 1817–1819, decorated locally by Henry Morris with a basket of flowers and a cobalt blue rim. Red stencil SWANSEA mark. Private collection

which received its Royal Charter in 1394, the Worshipful Company of Salters, formally ninth in the Guilds order, would have been very appropriate for this purpose since it involved chemicals manufacture and trade, which certainly embraced the raw materials of porcelain production. However, it was probably felt that the secrecy surrounding the manufacture of ceramics and the closely guarded formulations of the paste and glaze recipes defeated the purpose of open legislative control and the specification of standards, even though the quality controls exercised in the chemical preparation of the materials and in their firing were obviously appreciated as being essential for the achievement of a good commercial product.

References

J.A. Anderson, Derby Porcelain and the Early English Fine Ceramic Industry, Ph.D. Thesis, University of Leicester, UK, Oct 2000

H.G.M. Edwards, *Swansea and Nantgarw Porcelain: A Scientific Reappraisal* (Springer Publishing, Dordrecht, The Netherlands, 2017)

A. Vickery, *The Gentleman's Daughter: Women's Lives in Georgian England* (Yale University Press, New Haven, Connecticut, USA, 2004)

S. Whatman, *The Housekeeping Manual of Susannah Whatman* (National Trust Publishing, Pimlico Books, London, 1987)

Chapter 3
Analytical Results and Correlation with Recipes and Formulations

Abstract A survey is made of the raw materials which comprise the manufacture of Swansea and Nantgarw porcelains such as sand, potash, lime, flints and bone ash and the problems that occur in the interpretation of elemental data from the earlier gravimetric experiments in comparison with the more recent analytical instrumental results. A particular problem surfaces in the estimation of bone ash from the phosphorus content expressed as the pentoxide, phosphate and phosphoric acid. In addition, the variability in the imprecise way that the formulation of bone ash as a calcium hydroxy phosphate creates a separate issue here. Comprehensive data are collated from earlier and later analyses of Swansea and Nantgarw porcelains to illustrate the data obtained from the analyses performed to date.

Keywords Raw materials · Gravimetric analysis · Instrumental analysis
Bone ash · Correlation of data · Hydroxy apatite

3.1 Early Analytical Data Correlation with Composition

The much-admired Chinese porcelain, a *hard paste* and true kaolinic porcelain of clear translucency and whiteness, was described (Burton, *A History and Description of English Porcelain*, 1902) as "*a most highly developed form of true pottery ware, having clay and clay-forming minerals as its principal constituents*". As such, Chinese porcelain consisted essentially of china clay (*kaolin*, a plastic infusible aluminium silicate) and china stone (*petuntse*, a fusible mixture of alumina, potash and sodium silicate) which conferred the desired translucency upon the fired ceramic. In contrast, *soft paste* porcelain had a lower china stone and higher calcareous composition and early *glassy* forms also had a high phosphatic and silica content. The earliest comprehensive porcelain analyses of Eccles and Rackham in 1922 (*Analysed Specimens of English Porcelain*, 1922), following on from the analytical data of Sir Arthur Church on a much more limited range (Church, *English Porcelain*,

H. G. M. Edwards, *Nantgarw and Swansea Porcelains*,
https://doi.org/10.1007/978-3-319-77631-6_3

1894), were undertaken on marked or verified museum specimens from 18 different porcelain manufactories in the Victoria & Albert Museum Collection and gave results which expressed the comparative percentage compositions of idealised chemical components such as silica, alumina, phosphoric acid, soda, potash, lime and magnesia. These bulk analyses, in which compositional data were presented broadly in terms of metal or nonmetal oxides, could not reflect the true material and mineral composition of the initial porcelain body which had been chemically altered significantly after its processing in high temperature kilns during which several complex solid-state reactions had occurred. For example, Church (*English Porcelain*, 1894) in his description of Bow porcelain quoted 17.3% phosphoric acid, which he equated with a 43.8% bone ash content in the original paste, a material which has been incorrectly represented in the chemical formulations of most contemporary and subsequent authors in terms of a calcium: phosphorus ratio. Likewise, the presence of silica expressed as a component, SiO_2, does not take account of the silicate transformation of flint glass, feldspar and china clay at kiln temperatures into wollastonite, illite and whitlockite. Other specific analytical terms are also often rather imprecisely defined with regard to their chemical formulation: *potash* (from the ancient Dutch, *potasschen*, giving the older terminology *pot ash* for this material) should now be considered strictly to be chemically potassium carbonate, but in the 18th Century it was obtained by burning wood and leaching the ash with water, and finally reducing the aqueous extract to dryness—thus, natural "potash" contains potassium carbonate, potassium chloride and potassium sulfate, and caustic potash also contains potassium hydroxide. *Pearl ash*, cited by Lewis Dillwyn in several of his trial recipe formulations, is a granular and crystalline form of potassium carbonate obtained by extraction and crystallisation of these wood ash residues. *Soda* (from the Arabic *suwwad*) is an impure sodium carbonate obtained in a similar fashion from calcining specific woods and plants and can be a mixture of sodium carbonate (washing soda) and sodium bicarbonate (baking soda). A natural mineral equivalent of this is *natron*, which also contains sodium sulfate, carbonate, chloride and bicarbonate, and this was much used in ancient Egypt for the desiccation of human remains in their mummification rituals and the large supplies required of approximately 200 kilos per cadaver could be readily mined in the sedimentary deposits from the Western desert in the lakes of Wadi Natrun. Hence, the analytical results expressed for potash and soda represent the pure chemical extremes of potassium and sodium carbonates and not necessarily the real compositions of the potentially impure and perhaps imprecisely defined starting materials themselves however selective and rigorous the china manufactory proprietors were in their sourcing of these raw materials. The presence of significant quantities of impurities in these heterogeneous starting materials, which were often of indefinite chemical composition anyway, will naturally affect the true component compositions in the original paste and the quantitative analytical data derived from them as the latter will be expressed as pure chemical entities.

The respected 19th Century ceramic historian Robert Drane, who contributed a Foreword to William Turner's first classic text on Swansea and Nantgarw porcelains in 1897 (Turner, *The Ceramics of Swansea and Nantgarw*, 1897), commented that the chemical analysis of porcelains was not needed at all, since "*cats ignorant of analysis know good milk*" and that by analogy nothing could ever replace an expert opinion about the factory source of an unknown piece based on its applied decoration and stylistic grounds. In this, he was perhaps metaphorically overestimating one's own ability to unequivocally identify Swansea and Nantgarw porcelains from experience alone and, although this expertise is a most creditable and valuable accomplishment, this is a potentially dangerous statement because of the many clever forgeries that have been created and which circulate in the marketplace! Unfortunately, such a patently misleading remark has been supported by later authors, as exemplified by Jones and Joseph (*Swansea Porcelain*, 1988), and can perhaps be best translated as "*what one does not understand one does not really need to know*", thereby creating many pitfalls and reasons for the outright dismissal of analytical science in ceramics research. This apparent dismissal of scientific evidence and its relegation to second place behind expert opinion in the attribution of artworks is still a feature of many organisations and establishments and is exemplified by the statement expressed by Battie (*Antique Collecting*, 1994), an eminent and well-respected ceramic antiques expert and historian, who when asked where is the technology for ceramics analysis wrote:

> There isn't any .... Many tests are invasive-they cannot now, and probably never will, be conducted without damaging the object ...The ultimate test is the expert eye and a collector with perhaps only a few years' experience can still beat any technology available today. And probably always will.

This is indeed a condemnation of analytical science which perhaps was pretty accurate in 1994, but developments in instrumentation and application of analytical techniques over the past quarter century have now opened up some new possibilities as will be demonstrated here later, where expert opinion alone has been challenged and shown to be faulty and worthy of reconsideration regarding the attribution of otherwise exemplary pieces of porcelain.

How much more realistic and better appreciated it would be, therefore, for the recognition and the adoption of a *holistic* approach, as has been manifestly proved successful in forensic art science relating to paintings, whereby the scientific and art historical themes are considered together for the elucidation of problems associated with the attribution of unknown factories or specific artworks to different ateliers and for the establishment of ongoing parallel research themes in ceramics to elucidate the correct assignment of unusual pieces (John, *Nantgarw Porcelain*, 1948 and *Swansea Porcelain*, 1958; Edwards, *Swansea and Nantgarw Porcelains: A Scientific Reappraisal*, 2017). In defence of the adoption of an expert opinion solely to identify porcelain provenancing and sourcing, it must be stated here that the historical chemical analyses of porcelain pieces has invariably required their destruction completely or in part, unlike the analogous removal of micro samples of pigments from oil paintings to determine the paint usage chronology usually undertaken prior to their restoration. Hence, hitherto, chemical analysis in the ceramics field has been carried out on pieces that are already badly damaged (Eccles and Rackham, *Analysed Specimens of English Porcelain*, 1922) so that the taking of further samples and replicates from the specimens and effectively destroying them completely or in part is not seen to be problematic, as witnessed by the early work of Eccles and Rackham (1922) on specimens of Nantgarw and Swansea china (and specimens from Chelsea, Worcester, Derby and other factories) in the Schreiber Collection which were already broken or badly damaged. Likewise, the later work of Tite and Bimson (1991) and Owen et al. (1998 and *ff*.) used broken pieces or shards and one important piece of broken, decorated Swansea china from a named service in a museum collection (The Royal Institution of South Wales, Swansea). Hence, it is a major step forward scientifically in the current work to report chemical and molecular analytical data acquired from perfect, finished pieces of Swansea and Nantgarw porcelain without involving any mechanical or chemical pre-treatment or removal of samples from these in any way. To achieve this, Raman spectroscopy, a non-destructive light scattering technique using near-infrared laser excitation and a portable spectrometer with a fibre-optic cable to permit ease of access to specimens of unusual shape, was adopted to interrogate perfect pieces of porcelain from the Nantgarw and Swansea factories in a private collection to facilitate the acquisition of spectral data which can be assessed and evaluated for their characterisation and discrimination. In this way, it should be possible to determine the potential origin of unknown pieces of porcelain from the same time frame, ca. 1780–1830, spanning that of the Nantgarw and Swansea factory production years and in this same exercise, some "doubtful" porcelain attributions have also been tested. This same technique has been adopted successfully by Philippe Colomban to determine the composition of some early French porcelain, such as Sevres, Chantilly and Sceaux, and from initial experiments using destructive micro sampling, has progressed to non-destructive examination of porcelain bodies and glazes (Colomban 2004, 2005, 2013). To date, however the Raman spectroscopic technique has never been used to characterise Nantgarw and Swansea porcelains either destructively or non-destructively and this will be an objective in the present work. Hitherto, the only Raman spectroscopic study of British porcelain using this technique has been reported by Edwards et al. (2004) for a Rockingham porcelain inlay to a mahogany table made around 1835 and a comparison made of body components and pigments with a plate from a Rockingham porcelain service of a similar period, the latter marked with a red griffin mark.

3.2 Nantgarw and Swansea Porcelains: Statements for Verification

When one considers the detailed results of chemical analyses that have been carried out on Swansea and Nantgarw porcelains over the past century, it is also possible to identify several conclusions from which potentially incorrect or misleading statements have hitherto been made which can now be considered in a new light and to equally ascribe correct assertions based on definitive and quantitative determinations of chemical compositions of porcelain paste analyses. It is timely, therefore, to firstly re-evaluate the analytical chemical data pertaining to Welsh porcelains and to attempt to confirm or perhaps correct any potentially erroneous concepts, some of which are outlined below:

- Only one Nantgarw body was ever made by William Billingsley and Samuel Walker and this was later adopted at Swansea by Lewis Weston Dillwyn in 1814/15 to produce his esteemed duck-egg porcelain, using compositional details obtained from a secret recipe provided by William Billingsley and/or Samuel Walker, which was revealed only to Lewis Dillwyn at the Swansea China Works. Later analytical work of Owen and Morrison (1999) questions the assignment of only one body for Nantgarw porcelain and is strongly suggestive that Billingsley and Walker did experiment with variants and perhaps even introduced a commercial alternative which has not yet been recognised in existing specimens.
- From his Diary and Workbook (Appendix B), it appears that Dillwyn actually made up to sixteen distinctly different modifications (quoted by previous authors as a number between 9 and 12, but see Appendix B) to the Swansea porcelain body in his experimental trials with Samuel Walker in 1816/1817, but Eccles and Rackham (*Analysed Specimens of English Porcelain*, 1922) maintain from their analytical work that only three distinct Swansea paste bodies ever existed commercially, namely, the glassy, duck-egg and trident bodies, the former containing a significant component of glass cullet and the latter of soapstone or soaprock. However, later work by Owen et al. (1998) was strongly suggestive that a fourth porcelain body of a "hybrid" type and of a distinctly different silicaceous composition, containing a significantly high content of silica, was identified in several of the Swansea specimens they analysed.
- Despite statements to the contrary by Chaffers (*Marks and Monograms on Pottery and Porcelain with Historical Notes on Each Manufactory*, 1863) and Church (*English Porcelain*, 1894), it is now believed that there is no evidence that William Billingsley ever revealed the secrets of his porcelain composition to John Rose, who apparently purchased the moulds from the Nantgarw and Swansea factory closure sales in the early 1820s. It has always been maintained that Billingsley's Nantgarw porcelain recipe and compositional formulation died with him, but in 1885 a practising analytical chemist, Professor J. W. Mellor, succeeded in creating a new ceramic body which in all respects was an exact match for that of Billingsley's Nantgarw porcelain, from a formulation apparently revealed to John Taylor, author of The Complete Practical Potter (Shelton, 1847), by Samuel Walker personally. This comprised:

  > 26 lbs bone ash, 14 lbs Lynn sand and 2 lbs potash, mixed with water, this frit made into bricks and then fired in a biscuit kiln and ground up. 40 lbs of this frit was mixed with 20 lbs china clay for the final firing.

  It is indeed relevant that this is precisely the same recipe and composition used by Professor Mellor for his novel experiment to recreate Nantgarw porcelain, which after firing at 1300–1350 °C, followed by the appropriate glazing with a prescribed lead oxide slip, apparently reproduced the Nantgarw body perfectly in appearance and texture according to Nance (*The Pottery and Porcelain of Swansea and Nantgarw*, 1942) who confirmed this result by personal inspection of the synthetic porcelain produced in comparison with his specimens of genuine Nantgarw porcelain. Yet simple calculation reveals that the percentage of bone ash in this final body is 41%, which is significantly higher than the 33% bone ash content claimed originally for the Nantgarw secret recipe in other texts—giving a possible 20% error in formulation of this paste component alone!! As for the purchase and eventual removal of the Nantgarw and Swansea moulds by John Rose from the final auction sales there is no evidence that Coalport ever used these in the manufacture of their porcelain and perhaps they were acquired merely for study purposes or even to deny their acquisition by any other china manufacturer and competitor? A rather intriguing comment in the literature refers to an effort made by Samuel Walker to re-invigorate porcelain manufacture in Swansea or Nantgarw in about 1828, just after William Billingsley had died, but it seems that Lewis Dillwyn was no longer interested: this is relevant to the preceding statement that John Rose had removed all fabric of porcelain manufacture from Swansea in the early 1820s—if this had actually happened then Walker's idea would have been a clear non-starter as nothing would have been left to facilitate this enterprise!

- The Swansea trident body, according to the recipe formulated by Lewis Dillwyn and Samuel Walker and there identified in Dillwyn's workbook as Body Number 2 (Appendix B), comprised a frit of 8 lbs sand, 4.5 lbs soaprock (china stone) and 1.5 lbs potash which was then mixed with china clay as 5 parts to 1, respectively, contained no bone ash and possessed a very high silica composition (84%)—analytical figures therefore show effectively zero phosphorus content, lower percentages of calcium (1.0%), alumina (8.7%) and the presence of a significant amount of magnesium (2.5%) compared with the fabled "duck-egg" Swansea porcelain and its Nantgarw rival, where the silica percentage was approximately 47%, alumina 17%, phosphoric acid 15% and calcium oxide 15%.

3.3 Wet Chemical Analysis: Kiln Firing and Composition

The analytical data which have provided the source of all the early discourse about the composition of Swansea and Nantgarw porcelains were obtained from "wet chemical" analyses, during which a ceramic specimen was ground up and dissolved in mineral acids then treated according to established gravimetric procedures to determine quantitatively the composition of the key chemical components: these invariably were silica (SiO_2), magnesia (MgO), lime (CaO), potash (K_2CO_3), alumina (Al_2O_3), iron oxide (Fe_2O_3), titania (TiO_2), phosphoric acid (H_3PO_4), sulfate ions (SO_4^{2-}) and soda (Na_2CO_3). These data were derived from careful gravimetric experiments (Vogel, *A Textbook of Inorganic Quantitative Chemical Analysis*, 1961), replicated several times, with weighed precipitates collected after dissolution and separation procedures, followed by ignition of the precipitates and drying to a constant weight, and were presented as % age figures equated to a 100% total composition. However, it is known from daybook statements and work diary entries that other minor components added to the paste were present, such as borax and lead oxide, the former to aid fusion and the latter in the form of flint ground glass cullet. Generally, in such gravimetric chemical experiments the accuracy of determination of a component would be expected to be better than a fraction of several tenths of one percent, which would accrue to a cumulative total error of about 1% on a multi-component mixture analysis. An examination of the compositional data of Eccles and Rackham (*Analysed Specimens of English Porcelain*, 1922) for both Swansea and Nantgarw porcelains immediately exposes a problem in that there is a greater variance from sample to sample which cannot be attributed to significant experimental errors and the analytical precision alone. This will be discussed in comparative detail later, but a note should be made here that these variations in the analytical chemical data could well reflect either unrecorded changes in composition of the paste or perhaps a different source of the minerals used in its preparation. Dillwyn has made the comment that the presence of undesirable impurities or components in the raw materials can have a deleterious effect upon the appearance of the final porcelain body, such as iron (III) oxide and calcium carbonate, and as a result the selection of quality materials from different sources was an ongoing process and could well give rise to unexplained variations in minor elemental determinations derived analytically.

In this context, it is useful here to consider briefly the effect of the operating kiln conditions upon the paste components—which reflects upon some of the analytical difficulties raised earlier and informs our comparative reasoning and interpretations later. China clay, kaolin, formulated as $Al_2Si_2O_5(OH)_4$, and sourced in Cornwall for both Swansea and Nantgarw factory porcelain production, contains interstitial water which is expelled at temperatures between 500 and 600 °C, forming *metakaolin*, $2Al_2O_3.4SiO_2$. Further increase in temperature results in the evaporation of more highly coordinated water molecules until at 925 °C after an exothermic reaction a spinel is formed, $2Al_2O_3.3SiO_2$, which has resulted in the removal of a quarter of the silicon dioxide from the lattice and its conversion to cristobalite. Above 1050 °C a further reaction results in the formation of mullite, Al_2SiO_5. Meanwhile, around 1010–1100 °C feldspar, a potassium aluminium silicate formulated as $KAlSi_3O_8$, if present in the paste mixture composition will now dissolve the silicaceous cristobalite to produce a viscous liquid with corresponding decrease in viscosity at 1200–1250 °C, resulting in interstitial filling of the inorganic matrix and vacant coordination site occupation. At 1280 °C the mullite crystallises and the quartz remains dissolved in a very viscous material, the formation of hard paste porcelain has now been achieved. The addition of *petuntse* in Chinese porcelain manufacture served to provide a source for feldspar, quartz, and sericite, a fine grained micaceous hydrothermal alteration product of orthoclase and plagioclase minerals, formulated as $KAl_2(AlSi_3O_{10})(OH)_2$, which together with the kaolinite clay, was necessary for the sustainability of the high temperature fired Chinese porcelain, often produced in the Ming period Jinghdezen dragon kilns at temperatures in the region of 1350–1400 °C. In contrast, English soft paste porcelain required the addition of calcined bone ash and often glass cullet in place of the feldspar, which enabled a lower working kiln temperature and a high translucency to be achieved in the fired porcelain but often at the expense of shape retention of the artefacts arising from an increased thermal plasticity at higher temperatures—and resultant sagging or warping of the soft paste wares. This was the major factor that contributed to the demise especially of the operations at Nantgarw between 1815 and 1819. The further high temperature reactions that could arise between complex silicates due to the presence of phosphatic components in the added bone ash, such as calcium hydroxyapatite and whitlockite, and the formation of bytownite and wollastonite will be discussed later in the appropriate section. A very detailed study of kiln firing temperatures has been carried out by Owen and Morrison (1999) in their paper investigating sherds from the Nantgarw factory site, which they have described as sagged or unsagged, the term referring to the distortion of porcelain items occurring during the firing process which resulted in the extremely high kiln losses approaching 90%. Using the micro techniques of scanning electron microscopy coupled with energy dispersive X-ray spectroscopy, known as SEM/EDAXS, they were able to determine the mineral composition of the fired Nantgarw biscuit porcelain at the highest temperatures and the creation of the silicates and phosphates involved in the matrix. They demonstrated that at the highest temperatures involved in Billingsley and Walker's kiln, estimated to be in the range 1320–1430 °C, the porosity of the vitrified solid silicaceous matrix was filled with liquefied anorthite above the melting eutectic of the anorthite-tricalcium phosphate-silica system at 1300 °C and this would have accounted in large measure for the exceptional and noted translucency of the resultant Nantgarw porcelain by "filling the voids" with a glassy material. However, a negative consequence of this high temperature firing would also have been a decrease in the retention capability of the object's shape arising from an overfertile melt phase—which can be described as "overfiring". The key was accurate and precise control of the kiln temperature and firing times, which the Nantgarw personnel were unable to achieve: an idea of the magnitude of this kiln control problem can be gauged by an appreciation that a typical kiln may contain up to 20,000 items of porcelain being fired in a heating cycle for 40–60 h and then being retained in a cooling cycle of 40–60 h, both of these requiring temperature control to better than 20 °C at 1400 °C. The placement of batches of porcelain items at various positions in the kiln was also critical for the thermal efficiency and overall temperature control. In this context, the statement by Lewis Dillwyn in his Workbooks that several trial Swansea bodies utilised over 500 lbs of raw materials can be understood and it seems clear that several hundred if not thousands of items would have been fired in his each of his test experiments. We could therefore conjecture that, even if Dillwyn did not finally decide to utilise that particular paste for full commercial exploitation, he could well have instructed that these experimental items could have been decorated and dispersed locally: if some have survived they would therefore constitute some rather atypical Swansea porcelain compositions! The figures given by Owen and Morrison also could provide an idea of the enormous loss suffered by the Nantgarw operation—from a kiln firing of 20,000 items only 2000 would be judged fit for commercial sale—and this might even have been a whole week's factory production. William Billingsley's operation at Pinxton from his notes to John Coke, the china works proprietor, regarding raw materials costs had a rather more conservative production figure of approximately 70 tea sets per week in 1795—representing about 3000 items, and if this was a better estimate translated over to his Nantgarw factory production output then only 300 pieces of Nantgarw per week would have been saleable—perhaps a more realistic figure in view of the rarity of the surviving porcelain?

3.4 The Chemical Analysis of Porcelain

3.4.1 Early Data

The chemical procedures underpinning the gravimetric analyses are worth exploring here as they reveal potential problems which need to be accounted for in the interpretation of the results which hitherto have not been considered. Eccles and Rackham (*Analysed Specimens of English Porcelain*, 1922) and before them, Church (1894), would have used wet chemical methods to determine the quantitative composition of their porcelain specimens as at that time quantitative instrumental analytical techniques such as vibrational spectroscopy and spectrography were in their infancy. The data expressed in Church's book (*English Porcelain*, 1894) are derived solely from his analytical experimentation with Chelsea and Bow porcelains and he cites compositions from other factories which were possibly taken from published original recipes and formulations. Likewise, the established experimental procedures, which unfortunately were not stated specifically in Eccles and Rackham's booklet (Eccles and Rackham, *Analysed Specimens of English Porcelain*, 1922) summarising their analyses of English, Welsh and Chinese porcelains in the Victoria & Albert Museum Collection, would have involved the following standard analytical steps with only minor modifications:

1. The sampling of a specimen of verified factory provenance from the Victoria & Albert Museum Collection would have involved the taking of several grams or tens of grams from the object for replicate analyses to be performed. The damage caused to the ceramic artwork in this sampling process is destructive and non-reversible, as can be demonstrated by the existence of Eccles and Rackham's original specimens illustrated in their text and their specimen remnants which can still be viewed in the V&A Collection today and with photographs that are reproduced in Dr. Hillis' scholarly article (*Welsh Ceramics in Context II*, 2005, p. 170 and ff.). Tite and Bimson (1991) also draw attention to the "wet chemical" analytical procedures which Eccles and Rackham must have adopted to acquire their data: nevertheless, it has even been alleged elsewhere (Davis, *The Pottery Notebook of Maude Robinson*, 2007) that Eccles and Rackham categorically used spectroscopic instrumentation because that is what is used today—this is clearly a *non-sequitur*! From the relevant statement in the literature:

    > By the late 19th and early 20th century the analysis of ceramics by spectroscopic means was an important field of study. The culmination of these efforts is the classic Analysed Specimens of English Porcelain study by Herbert Eccles and Bernard Rackham published in 1922. (Davis 2007)

It is therefore patently incorrect and a comprehensive comparison between the earlier wet chemical analytical data of Eccles and Rackham and the modern instrumental analytical data forms an important part of this study.

2. The samples taken from the porcelain specimens for analysis were normally dissolved in a strong mineral acid: usually, hydrofluoric acid alone or in admixture with sulphuric or hydrochloric acids was used to convert insoluble silica and polymeric silicates into soluble fluorosilicates. An older method involved their thermal fusion with caustic alkali, which rendered the silicates chemically soluble in aqueous solution but left the silica as an insoluble precipitate. Following acid dissolution using an established procedure involving the addition of a sequence of chemical reagents used in quantitative analytical separation, elements of interest could be converted into insoluble precipitates of well-defined chemical composition and formulation, which could be filtered off and calcined or dried before carrying out their gravimetric determination. Then the quantitative composition of each element, such as calcium and magnesium, could be assessed according to the molecular formulation of its precipitate: silicon could be determined as silica, calcium as its oxalate or sulfate, magnesium as the pyrophosphate and potassium as its cobaltinitrite, from which the % ages of lime, magnesia, potash and other components could then be listed. The major issue here, of course, is that the representation and determination of these element percentages normally has no relevance at all to either the original mineral components in the porcelain manufacturing recipe or more importantly to the chemical constituents comprising the fired ceramic from the kiln after high temperature reactions in the kiln had occurred. For example, when Eccles and Rackham (Eccles and Rackham, *Analysed Specimens of English Porcelain*, 1922) cite the compositional data from their analyses in terms of lime and magnesia this does not mean that the porcelain specimen analysed after the firing process actually contained discrete chemical components of lime and magnesia, likewise for potash and soda. There must then be a back-calculation from the quantitative analytical data to interpret what the original components were and how much of each was used in the recipe. As can be appreciated, several assumptions have then to be made as to what comprised the original recipe for the porcelain paste and this may not be immediately apparent: for example, calcium occurs in calcined bone ash, calcite, dolomite and lime, whereas magnesium occurs in dolomite and feldspar, silicon in a variety of silicate minerals, clays, flints and sand, as well as glass frit, potassium in glass frit and in feldspar. The only unique elemental occurrences in most ceramic formulation components used in the early nineteenth century are lead in glass frit or cullet and phosphorus in bone ash: hence, previous writers have tended to describe porcelain compositions as low-phosphate or as highly phosphatic and lead-free or otherwise. However, the correlation of the percentage of phosphorus as determined analytically with an original % age of bone ash is not accomplished without some difficulty as it transpires that several errors have been made historically in the formulation of the derivative calcium hydroxyapatite assumed to be present in bone ash (Morgulis and Janacek 1931; Edwards, *Swansea and Nantgarw Porcelains: A Scientific Reappraisal*, 2017). This has necessitated a

correction factor to be applied here to enable comparisons to be made between the different analytical data for correlation with the original formulations. Also, the determination of lead content in porcelain bodies was often not carried out or reported, perhaps because of the perceived possibility of elemental interference from the surface glaze on the porcelain fragments or sherds which was commonly a lead oxide slip mixed with china clay and up to eight other components.

3. The basis of chemical gravimetric science applied to ceramics was perfected by Berzelius (Chalmers and Szabadvary 1980) in the early 19th century, building upon the careful studies of chemists such as Vauquelin, Klaproth and Davy: Davy's paper on the analysis of the mineral pigments in wall-painting fragments from Pompeii published in 1815 (Davy, *Philosophical Transactions of the Royal Society*, 1815) was a landmark in archaeological chemistry and it was not until 1921 when Baudoin published his analyses of bronze axe heads (Baudoin, *Comptes Rendus Hebdomadais de Seances de l'Academie de Sciences*, 1921) that spectrochemical procedures were successfully employed for archaeological analysis. Even so, it was only relatively recently that Richards in 1959 (Richards, *Archaeometry*, 1959) could publish the first spectrochemical analysis of archaeological ceramics, which has now progressively displaced the gravimetric determinations, resulting in novel information being forthcoming about the elemental and molecular compositions of ceramics and specifically porcelains through techniques such as XRF spectroscopy, SEM/EDAXS spectrometry, XRD diffractometry and Infrared and Raman spectroscopy. Modern instrumental analysis now provides micro-compositional data which facilitates the identification of the individual mineral phases, particles and mineral domain regimes in a porcelain body which cannot really be compared strictly with the bulk analytical data from the wet chemical analyses which preceded them in the early 20th Century. As mentioned above, the 19th and early 20th century analyses provided quantitative estimations of key metal oxides but the original materials in the fired porcelain from whence these were derived had to be deduced and this was often rather conjectural. For example, the estimated silica content measured from a calcined insoluble precipitate could have been derived originally from quartz, steatite, sand, glass, flints, serpentine and a variety of polymeric silicate networks, which therefore is not helpful for the assignment of a raw material used in the recipe for manufacture and subsequent back- tracking of the data into the original mineral compositions. Two recent papers by Pollard (2013, 2015) have given comprehensive accounts of the early analytical science of ceramics and the motivation for the adoption of chemical analysis which he has attributed to the desire of the Revolutionary government in France towards the end of the 18th Century to convert bronze church bells into cannon to further the military ambitions of Napoleon. A major problem for European ceramics production in the early 18th Century that we have alluded to earlier is that although supplies of the esteemed Chinese porcelain imports were readily available in Western Europe the manufacturers here had no idea of their chemical composition and source materials until later in the Century when the secret was brought from China by the French Jesuit missionary Father Francois Xavier d'Entrecolles.

Thus, there was a pressing need for accurate and quantitative chemical analytical input required to assist the formulation of European porcelain recipes to compete with these Chinese imports—this, therefore, was the major driving force behind the rapid adoption and development of the wet chemical analytical methods to these materials.

3.5 The Bone Ash Problem

It is clear from the rather disparate records that survive relating to the compositional recipes of the porcelain pastes used at Swansea and Nantgarw that bone ash was an important component in the manufacturing process. Basically, bone ash was derived from the high temperature calcination of good quality ox and cow bones at 1100 °C followed by grinding and milling the thermally treated product to a very fine powder: William Billingsley, in particular, set great store by the achievement of a very finely ground bone ash for the preparation of his Nantgarw frit, to which Lynn sand, potash and china clay were added, for which he used the services of a local corn miller who would then have presumably worked under his personal direction to achieve a satisfactory outcome. There is a record in the Derby correspondence between Joseph Lygo, their agent in London, and William Duesbury of the difficulty in acquiring good quality and consistently prepared bone ash, which was sourced from a variety of places, including the ex-Chelsea factory suppliers in London. Duesbury insists that his "strainer" for ground bone ash was a linen mesh of 1/150th inch porosity, corresponding to approximately 0.2 mm or 200 microns, but he emphasised that it was not always possible for his workforce to achieve this criterion of fineness. A detailed discussion of the potential problems associated with the achievement of a consistently finely ground component for porcelain manufacture is given in Sect. 4.4 of this text and an outline given of the potential contamination that could occur in this process. Although there are no records extant from the Nantgarw factory relating to the precise formulation of the recipe for its porcelain paste, Taylor (*The Complete Practical Potter*, Shelton, 1847) ascribes a recipe given to him by Samuel Walker, Billingsley's kiln manager, which comprised an initial mixture of 26 lbs ground bone ash, 14 lbs Lynn sand and 2 lbs potash mixed with water which was fired in a biscuit kiln. Then the 42 lbs of this cooled and finely ground frit were mixed with 20 lbs china clay to form the Nantgarw paste. Although this has been dismissed as hearsay or fanciful thought by some authors, on the 1st December, 1885, there appeared in the *Pottery Gazette* an account of experiments undertaken by Professor J. W. Mellor, a highly respected ceramics chemist from Stoke-upon-Trent, in which he described the creation of a porcelain body using this composition which matched very closely indeed the Nantgarw body in physical appearance and translucency (Nance, *The Pottery and Porcelain of Swansea and Nantgarw*, 1942; Turner, *Ceramics of Swansea and Nantgarw*, 1897; Edwards, *Swansea and Nantgarw Porcelains: A Scientific Reappraisal*, 2017). Although it is generally believed that William Billingsley never revealed his formulation recipe personally for Nantgarw porcelain to Lewis Dillwyn at Swansea in 1814 or to John Rose at Coalport in 1820, it is nevertheless quite possible that after Billingsley's death on January 16th, 1828, Samuel Walker, who would have been privy to the final formulation used at Nantgarw having been involved for many years previously in Billingsley's experiments to perfect the porcelain body, could have released the recipe to other interested sources before he departed to set up his earthenware factory in New Jersey, USA. Samuel Walker remained at Coalport "to improve the art of china making" (Haslem, *The Old Derby China Works*, 1876) but departed thence later in 1828. After leaving Coalport he severed all his links with porcelain manufacture in Britain—particularly following an allegedly apparently unsuccessful attempt to re-start production first at Swansea with Lewis Dillwyn and then at Nantgarw with the successors of Thomas Pardoe around 1830. Thereafter he seems to have disappeared from the records in the following 14 years. Samuel Walker emigrated to the United States of America, arriving in New York on the 22nd April, 1842, with a new family, comprising his second wife, also called Sarah (who was born in 1815) and a daughter Mariah (born in 1834) where he manufactured pottery and brown glazed earthenware of the "Rockingham type" in an earthenware ceramics manufactory in Temperance Hill, West Troy, New Jersey, where he died at 90 years of age, being pre-deceased by Sarah in 1862 (Broderick, *An English Porcelain Maker in West Troy*, 1988; Barker, *Pottery and Porcelain in the United States*, 1893). It is intriguing that a recent excavation of the factory site in New Jersey has discovered porcelain shards, which could suggest that Walker had also embarked upon porcelain manufacture there, but finished pieces of porcelain attributed to this source have not been identified. His release of specific information relating to the Nantgarw paste formulation could well have coincided with his emigration and start-up of a new life in the United States of America, without any punitive response emanating from the Royal Worcester Porcelain Works as had threatened him and Billingsley when they had departed from their original employment in Worcester with Martin Barr in 1810. Barr had died in 1812 and his successors at Worcester then made a decision to revert to the manufacture of more standard porcelain and to cease their experiments to achieve a novel porcelain body. This recipe recounted in *The Complete Practical Potter* (Taylor 1847), in fact, is the only note that survives that gives us any indication of the composition of the Nantgarw paste, which hitherto was thought by some to have died with Billingsley who had been noted for his secrecy in this respect. Although there is still uncertainty as to the precise kiln firing parameters used at Nantgarw, despite the comprehensive research undertaken by Owen and Morrison (1999), Professor Mellor's successful experiments in the synthesis of a Nantgarw analogue porcelain body in 1885 are indicative that he must have achieved very similar operational procedures in timing and kiln temperatures to those actually used originally at Nantgarw by Billingsley and Walker to produce a porcelain that was so similar to the original version in texture and translucency. For the first time we can estimate the correct % age of bone ash used in the Nantgarw formulation: this can be calculated as 26 lbs in a total paste mixture of 62 lbs, therefore equating to 42% bone ash, whereas previous literature had "estimated" a composition ranging anywhere from between about 50–75% bone ash in Nantgarw porcelain without a citation of any evidence at all for this quantitative compositional value. This provides another example of how a simple guess initially, however educated and informed, can be transcribed into a factual datum from which erroneous conclusions can then be made in future quantitative calculations of porcelain composition.

At Swansea, there are in existence detailed handwritten notes from Lewis Dillwyn's experiments undertaken with Samuel Walker and William Weston Young from 1815 to 1817 (see later and in Appendix B), in which he gives a recipe comprising 24 parts of bone ash mixed with 8 parts of flint, 16 parts of St Stephen's clay, 5 parts of Norden clay and 1 part smalt which gives "*a beautiful white opaque body—my finest produced at Swansea*" (Lewis Dillwyn, *Notebooks*, Appendix B) and must therefore surely refer to his prized successful effort in the production of his celebrated finest duck-egg porcelain. This formulation composition equates to 44.4% bone ash, which is remarkably similar to the corresponding figure which we have deduced comprises the Nantgarw formulaic recipe from Professor Mellor's experiments, viz. 43%. Perhaps this correlation has given rise to the historically erroneous statement that William Billingsley on his employment with Dillwyn to make the first Swansea porcelain in 1814 divulged the actual Nantgarw recipe to Dillwyn and that this was used in the Swansea China Works thereafter. However, there is documentary evidence that Dillwyn was seriously perturbed by the receipt of a letter from Messrs. Barr, Flight & Barr at Worcester, firstly cautioning him against employing Billingsley and Walker at Swansea and then further threatening him with legal action if he used Billingsley's porcelain recipe to make china there. This letter has been the subject of much conjecture and analysis in itself, since it is also known that Billingsley and Walker left their employment amicably with Barr, Flight & Barr at Worcester and were even presented with a parting gift of £250 at their severance, with which they commenced the start-up of porcelain manufacture at Nantgarw in the first phase in 1811/12. In the intervening period when manufacture had firmly commenced at Nantgarw, Billingsley and Walker then made an unsuccessful application for support funding, resulting in the closure of this enterprise on financial grounds. During this time no challenges or threats were forthcoming from Messrs. Barr, Flight & Barr at Worcester to Billingsley and Walker whilst they were operating at Nantgarw, so it can be concluded that it was the link-up and perceived danger of competition from Billingsley and Walker with the financial prowess of Lewis Dillwyn for novel porcelain manufacture at the Swansea China Works that prompted this course of action by the Worcester owners (Edwards, *Swansea and Nantgarw Porcelains: A Scientific Reappraisal*, 2017).

3.6 Glass Frit

An interesting point now emerges from a consideration of these recipes and associated analytical data in that it can be categorically stated that the Nantgarw porcelain formulation did not contain any glass frit: despite this clear fact, several authors have attributed the remarkably superior translucency of Nantgarw porcelain directly to a high content of glass frit in the paste, which would have been believed to have materially assisted in the clarity of the melt and its subsequent solidification. This possibly arose from the early experiments at Chelsea in the 1750s where glass frit or cullet was incorporated into the porcelain paste and incomplete mixing resulted in the appearance of a characteristic molten phase of "moons" of exceptional translucency which could be viewed in early Chelsea porcelain bodies by transmitted light. This cannot be the case here, since the recipe proposed by John Taylor which he reportedly received from Samuel Walker does not include a glass frit component at all—the precise analytical data of the 1990s support this conclusion since the lead content of Nantgarw porcelain is effectively zero, whereas flint glass frit would have contributed to a significant presence of lead because of its significant lead (II) oxide content. The lead content in flint glass is variable and can be as little as 5% or as high as 60%, with several authors elsewhere claiming that porcelains contained lead glass frit with 20–30% lead content. It is true that Eccles and Rackham (*Analysed Specimens of English Porcelain*, 1922) have reported small quantities of lead in their data for several porcelains that they studied, but it should be remembered that their analyses were carried out on finished, glazed articles which had been reduced to powders upon grinding (which would have incorporated both the glaze and the porcelain body—and also, of course, any associated applied enamels or gilding thereon) and most glazes at that time also contained a lead (II) oxide additive. A comment made by Sir Arthur Church (*English Porcelain*, 1894) in his analysis of the sole Brancas-Lauraguais porcelain specimen recovered from the fire- damaged exemplars at the Alexandra Palace in 1873 specifically relates to his "grinding off" the glaze from the fragment concerned before subjecting it to dissolution and chemical analysis.

It is relevant to note here that John Rose won a prestigious award in 1820 for a patent glaze on his Coalport porcelain which substituted cassiterite (tin oxide) for lead oxide, so reducing the toxicity of lead in the workplace- at that time the harmful effects of lead poisoning in the ceramics industry were becoming well-known. Dillwyn's recipe for his best duck-egg china on the other hand contains 8 parts in 52 of the body formulation of flint glass frit, corresponding to 14.8% glass content, and of this some 5% at least will be lead content and perhaps more. Actually, the situation is a little more complex than at first seemed because Dillwyn specifically mentions that he also added 1 part of smalt to his porcelain paste—and smalt, also known as cobalt blue, is a blue glassy material, analysing as cobalt aluminosilicate. Hence, the true glassy content of Swansea duck-egg porcelain is thereby increased to 16.7%, but not all of this would then be lead-containing material. It is even more interesting when we consider some of Dillwyn's other favoured recipes from the trial experiments that he carried out with Samuel Walker: Dillwyn's porcelain paste, which he estimated would rival the finest Chinese eggshell porcelain, comprised only 16.0% bone ash, 8 parts in 50 parts of mix, and no glass frit at all! A sequence of alternative formulations contained china clay, soapstone and bone ash only, giving a "*trident body*" classification, contained 33.0, 35.0 and 37.5% bone ash: this body was described by Lewis Dillwyn as an "*excellent body … a better improved body … and a harder body …*" respectively, but as we have noted above, the increased desired robustness of the china body could only be achieved at the expense of the texture and translucency of the finished article.

It is conceivable that the word "*frit*" as used in porcelain nomenclature in notes and recipes has been misinterpreted hitherto by some authors: in paste recipes, the china clay, bone ash and other components are usually premixed and then fired, upon which they are then cooled and finely ground before the addition of other components is made—this process is called *fritting* and the mixture is known as a *frit*. When flint glass is added, it is ground and this also then becomes a *frit*, so the opportunity for confusion in the terminology is apparent especially when one reads the original rather terse notes in work books which accompany the observations of the empirical experimentation involving several changes in paste composition. The failure to detect lead in the porcelain analyses can then directly be ascribed to the absence of flint glass frit (which is high in lead): another possible explanation for the absence of lead in the analyses could be that a soda glass or crown glass cullet has been used as a lower melting glass frit, in which case it would be lead-free but the data would then also reveal significantly higher percentages of sodium and potassium in the analyses. However, the elemental constituents of *pearl ash* added to the paste as a flux by several manufactories also contain sodium and potassium—quoted, for example, as *soda* and *potash* by Eccles and Rackham (*Analysed Specimens of English Porcelain*, 1922). Other factories which are contemporaneous with Swansea and Nantgarw used different paste compositions—and this, of course, can be considered objectively from a diagnostic viewpoint in the forensic characterisation and attribution of unknown porcelains to specific factories. For example, Owen and Barkla (1997) have reported their analyses of Derby porcelain from the first quarter of the 19th Century and have found two distinct types of body: a glassy porcelain containing 30–35% quartz, 34–56% flint glass frit, 6–10% china clay, 16% calcite and 6–10% bone ash, whereas a contemporaneous phosphatic type of porcelain contained 31–43% bone ash, 21–24% china clay, 31–39% china stone and 1–6% calcite. The variability of the components in the paste composition is a matter of record and explains why quantitative analyses of specified additives alone cannot be used as a discriminator in porcelain and ceramics identification. Although both recipes contain calcite and quartz, wollastonite is found only in the glassy formulation because in the phosphatic version all calcite is consumed by the formation of bytownite in a reaction between calcite and kaolin to form plagioclase at temperatures in excess of 200 °C. These compositions for Derby porcelain are significantly different to those found in the recipes discussed above for Swansea and Nantgarw china. A further analysis of a piece of later Bloor Derby porcelain (Owen and Barkla 1997)—comprising a biscuit shard of Queen Victoria in her coronation robes dated 1837—yielded the figures of 40.6% bone ash, 34.3% kaolin, 21.6% quartz and 3.5% calcite, again significantly different from the earlier phosphatic china stone porcelain body and perhaps containing glass frit from the high quartz content, although this has not been specifically mentioned in any recipe from that time.

3.7 Nantgarw and Swansea Porcelain: The Phosphate Enigma

If we now move to consider the analytical data from Nantgarw and Swansea porcelains in detail it is found that some recalculation is necessary to correctly estimate the % age of bone ash and other components, upon which much interpretation has been forthcoming, not all of it seemingly correct. It has been stated earlier that the major problem arising from the quantitative wet chemical analyses undertaken earlier is that the % age composition of the porcelains analysed is given for the elemental oxides seen as being most appropriate to the substance analysed following the breakdown of the polymeric ceramic matrix and these may have little relevance to the actual chemical entity present in the fired porcelain itself. For example, silica is quoted as SiO_2 and is derived from a variety of mineral sources in the porcelain matrix, which are all of different formulation chemically: sand, china clay, quartz, smalt and flint or soda glass all contain silica units as major components in their chemical structures, from SiO_2 itself to anionic species such as SiO_3^{2-} and SiO_4^{2-}, through to complex three-dimensional chains and matrices involving -O-Si-O- units. Lime, calcium oxide CaO, is quoted as a component in early analyses but the metallic calcium arises from its elemental presence in clay, calcite, bone ash and smalt, wherein it actually occurs in several different and unrelated chemical species such as calcium carbonate, calcium silicates and calcium hydroxyapatite. The presence of phosphorus is quoted in analyses as one of three possible formulations of phosphorus (V) chemically: namely, phosphorus pentoxide, P_2O_5, phosphoric acid, H_3PO_4, and phosphate, PO_4^{3-}, which have then been correlated with the % age of bone ash in the porcelain paste. Herein, lies a major problem for analysts since over the ages the chemical composition of bone ash has been debated and the incorrect formulation has frequently been applied; this can have serious consequences for the derivation of bone ash content in porcelains, especially where the corresponding records or precise formulation recipes are sparse or non-existent, as occurs for several important factories including the two under discussion here, Swansea and Nantgarw.

Table 3.1 gives the collected quantitative analytical data from the literature on Nantgarw porcelain and Table 3.2 the analogous data are presented for Swansea porcelain. Also, in no analytical literature to date relating to the Nantgarw and Swansea factories does there appear a precise statement as to the basis for conversion for the estimation of bone ash content derived from the phosphorus elemental determinations: hence, it is necessary to now attribute a "conversion factor" to make a comparison between the estimations of bone ash content expressed by different analysts. It is of relevance here to compare the early analyses of phosphatic Bow porcelains dating from 1750–1760 in the Victoria & Albert Museum collections first cited by Eccles and Rackham (*Analysed Specimens of English Porcelain*, 1922) with the later analyses reported by Tite and Bimson (1991) and the most recent of Ramsay and Ramsay (2007a, b). These data are collected in Table 3.3 and of particular interest is the comparison between the early gravimetric determinations of Eccles and Rackham and the instrumental analyses using the SEM/EDAXS equipment of Tite and Bimson, who took powdered specimens of the same specimens, and Ramsay and Ramsay, who undertook their measurements on intact or broken pieces of the same specimens, so it is now possible to compare the effect of sampling regimes as well. The first point to note is that these analytical data determined from a variety of gravimetric and instrumental methods over a period of 80 years is remarkably well-correlated: the instrumental methods have detected lower-level components such as iron, sulfur and titanium oxides. Of these, the iron oxide source can be attributed to the sand used in the recipe formulation, the titanium oxide to the kaolin (china clay) and the sulfur oxide to the gypsum or alabaster components. A point of contention must be made regarding the direct transposition of the Eccles and Rackham phosphorus determination as phosphoric acid, H_3PO_4, for comparison with the other two determinations as phosphorus pentoxide, P_2O_5, when we have already calculated that this would not be feasible without the application of a conversion factor of approximately 0.72 to compensate for the greater % age of phosphorus in the latter compound. Hence, the real comparator figures for Eccles and Rackham's determination should now be 13.65 and 9.83%, respectively, for Specimens 1 and 2, which provide a greater discrepancy with the values obtained from the later analytical results of Tite and Bimson and Ramsay and Ramsay. As Dr. Hedges has commented (in Twitchett, *Derby Porcelain 1748–1842*, 2002) this could reflect the greater analytical uncertainty in determination of phosphorus by the wet chemical method and we should therefore not realistically expect such a good agreement between the analytical results obtained using gravimetric and instrumental techniques for this element. The phosphorus oxide determinations, when converted into bone ash % age component, therefore give estimates of 47 and 40% for bone ash content in Specimens 1 and 2, respectively, for Bow porcelain in the 1750–1760 period.

In this context, the variability in composition of "bone ash" itself needs to be considered: the first reference to the adoption of bone ash in porcelains is found in a patent taken out by the Bow factory in 1746, under the terminology of "*virgin earth*", which appeared to be a mixture of bone ash, glass cullet, gypsum and alum—a later version of this material used in the period 1755–1769 was simplified and comprised only 90% bone ash and 10% gypsum. The glass cullet used in the earlier composite was not specified but believed to be an indefinite mixture of lead-free soda glass and lead-containing flint glass (Ramsay and Ramsay 2007a, b, 2017). Owen (2002) first commented upon the variability in composition of bone ash arising from differences in formulation: bones calcined below 1000 °C could be represented by calcium hydroxyapatite of formula $Ca_3(PO_4)_2.Ca(OH)_2$, giving a typical analysis of CaO 58.37%, P_2O_5 36.94% and H_2O 4.69%, which equates with a CaO: P_2O_5 ratio of 4.00, whereas when fired to temperatures approaching 1400 °C further dehydration to whitlockite occurs, a beta-calcium triphosphate of formula $Ca_3(PO_4)_2$, which equates to a rather different CaO:P_2O_5 ratio of 3.00. In their paper on Nantgarw porcelain shards, Owen and Morrison (1999) also propose that in the presence of other components in the kiln a partial transfer of calcium occurs in the low-fired calcium hydroxyapatite through the volatility of calcium oxide as well as loss of water

Table 3.1 Analytical data for Nantgarw porcelain (data expressed in % ages)

Analyst	Specimen	SiO_2	Al_2O_3	P_2O_5**	CaO	MgO	Fe_2O_3	K_2CO_3	Na_2CO_3	PbO	SO_4^{2-}	Equiv. BA (%)
Eccles and Rackham (1922)	*NG21	46.0	17.0	13.9	19.7			2.7	0.4			35
	*NG22/*Duncombe*	38.9	18.1	17.1	22.5			2.6	0.1			40
Tite and Bimson (1991)	NG14/1	43.8	12.5	17.4	22.5	0.6		2.5	0.8			
	NG18/4	45.0	13.3	16.4	21.2	0.5	0.4	2.2	1.0			
	NG18/7	44.6	13.3	16.4	21.9	0.7	0.2	2.2	0.7			
Owen et al. (1998)	NG8	43.5	12.7	17.6	23.3	0.4	0.2	2.3	0.4			
	NG/N2	70.8	8.9	7.5	9.9	0.2	0.2	2.3	0.2			
	NG/N4	80.3	9.1	0.5	0.6	1.1	0.1	5.6	1.8			
Owen and Morrison (1999)	N11	45.7	13.8	15.4	21.4	0.5	0.2	1.7	0.5		0.7	
	N12	46.1	14.0	15.0	21.0	0.5	0.2	1.8	0.5		0.8	
	N13	43.7	12.7	17.1	22.9	0.4	0.2	2.4	0.4		0.2	
	N14	42.4	13.4	17.4	23.7	0.4	0.2	1.9	0.4		0.2	
	N23	39.0	13.3	19.1	25.5	0.4	0.2	2.0	0.4		0.1	

(continued)

Table 3.1 (continued)

Analyst	Specimen	SiO_2	Al_2O_3	P_2O_5**	CaO	MgO	Fe_2O_3	K_2CO_3	Na_2CO_3	PbO	SO_4^{2-}	Equiv. BA (%)
	N24	41.1	13.3	17.8	24.3	0.4	0.2	2.3	0.5		0.1	
	*N37	45.5	12.1	16.3	21.8	0.4	0.2	2.9	0.6		0.1	
	N25	80.7	9.0	0.5	0.5	2.2	0.1	5.4	1.5		0.1	
	N43	80.5	9.3	0.4	0.4	2.2	0.1	5.4	1.5		0.2	
	N44	80.3	9.3	0.5	0.6	1.8	0.2	5.5	1.5		0.1	
	N34	78.0	6.6		0.6	1.0		12.2	1.6			

** refers to the conversion of values of phosphate

Notes

1. Of the nineteen specimens of Nantgarw porcelain analysed and cited in this table only three, indicated by a *, are from finished and decorated pieces, all from Museum collections
2. Data expressed variously as H_3PO_4 (Eccles and Rackham), P_2O_5 (Tite and Bimson; Owen and Morrison) and PO_4^{3-} (Owen et al.) in original reports: all have been converted to P_2O_5 figures to facilitate comparison using conversion factors described in the text a **
3. Separate analyses of the glaze on the silicious sherds N25, N43 and N44 and of the specimen N37 (finished porcelain) give % PbO compositions as 15.4, 14.7,14.4 and 22.6%, respectively, with an average SiO_2 content of 63.9 ± 6.0% and Al_2O_3 of 7.5% ± 1.5%
4. The Nantgarw porcelain body contains no PbO, confirming that flint/lead glass cullet was not a component in the porcelain body manufacture

Table 3.2 Analytical data for Swansea porcelain (data expressed in % ages)

Analyst	Specimen	SiO_2	Al_2O_3	$P_2O_5^b$	CaO	MgO	Fe_2O_3	K_2CO_3	Na_2CO_3	SO_4	TiO_2	Equiv. BA (%)	SS (%)
GVLE 1914 (Eccles and Rackham 1922)	GVLE	44.3	21	12.9	18.1	0.7	1.9	1.0				33	
[c]Eccles and Rackham (1922)	SW23	47.8	26.5	9.9	13.3		3.1	0.2				25	
	SW40	84.0	8.7		1.0	2.5	2.9	0.6					7.5
	SW41	81.6	8.9	0.3	0.7	4.3	3.3	0.6					1.3
Owen et al. (1998)	S1	68.0	8.2		9.7	0.2	0.1	2.5	10.9	0.2			
	S2	71.1	24.9	0.2	0.3	0.2	0.7	1.6	0.4	0.2	0.4		
	S3	73.8	21.6	0.2	0.3	0.2	0.8	1.6	0.6	0.4	0.4		
	S4	70.7	24.5	0.1	0.4	0.2	0.7	1.8	0.7	0.3	0.4		
	S5	70.5	25.1	0.2	0.3	0.2	0.7	1.7	0.5	0.3	0.4		
	S6	71.7	23.9	0.2	0.2	0.2	0.6	1.9	0.6	0.3	0.3		
	S7	73.6	22.1	0.3	0.3	0.2	0.6	1.7	0.6	0.3	0.4		
	S8	70.0	25.4	0.2	0.4	0.2	0.7	1.7	0.5	0.4	0.5		
	[a]S9	45.2	20.0	13	16.7	0.4	0.3	2.8	1.0	0.1			

[a]Biddulph service
[b]Expressed variously as H_3PO_4 (E and R), P_2O_5 (T and B), PO_4^{3-} (O et al.)
[c]*GVLE* Sample from Loan Exhibition Glynn Vivian Art Gallery

Table 3.3 Comparison of analytical data (%) for phosphatic Bow Porcelains, 1750–1760

Oxide	Specimen 1: Sauceboat, C673–1920			Specimen 2: Fruit Dish C16–1920		
	[a]E and R (1922)	T and B(1991)	R and R(2002)	E and R (1922)	T and B(1991)	R and R(2002)
SiO_2	43.58	45.6	46.4	50.38	49.2	50.31
TiO_2	–	0.5	0.3	–	0.3	0.3
Al_2O_3	8.36	8.7	8.5	5.78	5.6	5.3
Fe_2O_3	–	0.5	0.4	–	0.3	0.3
MgO	0.6	0.6	0.4	trace	0.3	0.3
CaO	24.47	23.6	23	24.87	24.5	22.87
Na_2O	1.2	0.8	67	0.7	0.5	0.85
K_2O	0.85	1.1	0.95	0.53	0.6	0.55
P_2O_5	18.95	18.6	19.4	13.66	16.2	16.28
PbO	1.75	–	–	1.49	0.4	0.2
SO_2	–	–	–	–	2.1	2.75

[a]Eccles and Rackham (1922)
Tite and Bimson (1991)
Ramsay and Ramsay (2007a, b)

and at intermediate temperatures between 1100 and 1400 °C ratios of $CaO:P_2O_5$ of 3.2–3.8 can be anticipated.

Finally, Twitchett (*Derby Porcelain 1748–1842*, 2002) and Owen (2002) both comment independently on the analytical difficulties inherent in the determination of elemental phosphorus, particularly in the early gravimetric wet chemical analytical work on ceramics: John Twitchett cites Dr. Robert Hedges' commentary on the analysis of Derby china in that it is not possible to determine the extent of bone ash incorporated since the techniques involved cannot measure phosphorus with sufficient sensitivity, but that inference as to the bone ash content could be made through an examination of the K:Ca ratio, for which the percentage of Ca is increased through bone ash addition.

3.8 Comparative Analytical Data

The following points can be made in an objective assessment of the analytical data that currently exist for Swansea and Nantgarw porcelains and which have been collected in Tables 3.1 and 3.2:

- Only a relatively small number of 29 items of Nantgarw porcelain and 13 items of Swansea porcelain have ever been analysed chemically hitherto and it should be noted that only some of these results are reported in the open literature: of these, four items can be identified (Edwards, *Swansea and Nantgarw Porcelains:*

A Scientific Reappraisal, 2017) as being from *named services*, for which supportive historical provenancing data can be accessed. Sample NG22 (Victoria & Albert Museum Collection) is a saucer from the Nantgarw tea service originally made for Mr. Edmund Edwards who was landlord of the Nantgarw House, Tyla Gwyn, rented by William Billingsley and adjacent to the China Works site. This was decorated by Billingsley personally at Nantgarw and was one of four services given to Edmund Edwards *in lieu* of rent; this service passed through family inheritance to the Rev. Duncombe of Llandaff, before being dispersed eventually at auction—hence its acquisition of the associated name, the *Duncombe* service. A photograph of the actual Nantgarw saucer analysed by Eccles and Rackham in 1922 is reproduced in Dr. Hillis' article (*Welsh Ceramics in Context I*, 2005, page 184 and Colour Plate 9.17) where it can immediately be identified as part of the *Duncombe* service (although this fact was not realised or stated by Eccles and Rackham at the time of their analytical study). Sample SW/S9 (obtained from The Royal Institution of South Wales, Swansea) is a fragment from an oval platter from the *Biddulph* service, a 100-piece dinner-dessert service ordered by Lord Biddulph of Ledbury Park and decorated with landscape scenes of his estate and surrounding countryside in the London atelier of John Bradley by Philip Ballard. In addition, sample SW23 (Victoria & Albert Museum Collection) is a fruit comport from a Swansea dessert service which is of an identical pattern to the Nantgarw *Hensol Castle* large breakfast—dessert service of over 300 pieces commissioned by Lord Crawshay of Cyfarthfa Castle, whose family line inherited Hensol Castle from Baron Talbot. A fourth example, analysed by Herbert Eccles in 1914 from the Glynn-Vivian Loan Exhibition (GVLE), organised on the occasion of the centenary of foundation of porcelain manufacture at Swansea, is a blue and white transfer printed porcelain dish from the *Gibbins* service which is somewhat unusual in that it bears an impressed BEVINGTON & SON mark in addition to the red script stencilled SWANSEA mark. This piece is not of the usual textural quality expected for duck-egg porcelain and the analytical data shown in Table 3.2 are strongly suggestive that this was an example of an experimental porcelain body that possibly was discarded initially as not being satisfactory but then perhaps decorated and marketed for sale at the closure of the Swansea factory.

- The importance of named services with their associated historical provenancing is that they are documentary of a particular period of production in the factory: this can be vital information in assessing the influence of later compositional changes effected by empirical experimentation, which as we have seen above was certainly ongoing in attempts to improve the paste characteristics. One must still be rather careful in making the assumption that because a piece is part of an identified named service then that must imply that it indisputably originates from that particular factory: there are several instances recorded where the Swansea China Works, for example, or their agents most probably "imported" items from competitors' factories to make up large commissions where kiln failures or other operational reasons resulted in their inability to otherwise meet a chronological deadline for the provision of a commissioned service. Examples include the *Lysaght* service, *Gosford Castle* service and the *Biddulph* service mentioned here.

- Of these analysed samples of Nantgarw and Swansea porcelains, the two Nantgarw specimens analysed by Eccles and Rackham (*Analysed Specimens of English Porcelain*, 1922) were in the Victoria & Albert Museum Collection, three shards of glazed porcelain from unspecified locations at an excavation at the Nantgarw site in 1932 were obtained from the British Museum archive and analysed by Tite and Bimson in 1991, and Owen et al. in 1998 acquired 10 shards from the Nantgarw factory site and 8 from the Swansea site, a ninth Swansea example coming from a specimen held at the Royal Institution of South Wales; of the latter, only one Nantgarw shard was marked and no Swansea shards were marked. In a later paper, Owen and Morrison (1999) analysed ten more shards from the Nantgarw site, which had not appeared in the earlier paper, but they also then had access to a florally-decorated specimen plate of finished Nantgarw porcelain from the Nantgarw China Works Museum, labelled as N37, from which they were able to give analytical data from the porcelain body and the glaze by sampling the unglazed footrim. An unusual glass shard from the Nantgarw site was also analysed as being a suspected sample of Billingsley's vitrified glaze, but this could not be verified from its composition, which differed from that provided as the typical Billingsley glaze as stated to Taylor in 1847 in *The Complete Practical Potter*. Whereas Eccles and Rackham (*Analysed Specimens of English Porcelain*, 1922) were able to analyse finished porcelain pieces from the Victoria & Albert Museum collection (comprising 2 Nantgarw and 3 Swansea finished and decorated specimens), all other studies have been carried out on shards recovered from the vicinity of the porcelain manufactories, with the sole exception of the *Biddulph* (ex-Royal Institution of South Wales, Swansea) and *Gibbins* (ex-Glynn Vivian Gallery Loan Exhibition, Swansea) Swansea service items reported above and the most recent example cited by Owen and Morrison in 1999 of Nantgarw porcelain(ex-Nantgarw China Works Museum). It is recognised that to date the main waste dumps belonging to the Nantgarw and Swansea factories have not been located specifically, although several large deposits of broken china have been found scattered at the sites, so this immediately raises the question as to the authenticity of origin the shards that were found "*lying around*" on site (Owen et al. 1998). Only one shard had a factory mark and in most cases, therefore, a matching of shard shape with items of known factory output would be essential for verification of this attribution. Owen and Morrison (1999) draw attention to this perhaps dubious provenancing of the shards in their analytical data conclusions and the as yet unfulfilled discovery of the actual waste tips at both factories. The stratigraphic archaeological excavation of fragments is critical for placement of the finds in a chronological context and in the particular cases cited here for Swansea and Nantgarw would be essential to place the shards of unusual chemical composition in a correct temporal relationship or timeline which runs concurrently with any developmental work or studies made on paste improvements. Then, the rather strange, highly silicious sherds identified by Owen and Morrison in their analyses could really be examined in a chronological context and correlated with any experiments being made by Billingsley and Walker whilst they attempted perhaps to improve the survivability of their porcelain in the kilns at the extremely high temperatures adopted for firing.

- Current thinking is that Billingsley and Walker made just one Nantgarw body and this is accepted in the absence of any documentation hinting otherwise: however, Owen and Morrison's work might be strongly suggestive of an alternative scenario which points to some unsuccessful experiments having been undertaken by Billingsley and Walker and the possibility that a second Nantgarw body does exist, which has not yet been identified analytically in perfect finished porcelain pieces. In addition, it is acknowledged that for the completion of large commissions, especially accounting for the exceptionally large kiln wastages at Nantgarw and Swansea which rendered perhaps at most only 10–20% at most of fired porcelain immediately saleable, there is a distinct possibility that several items would have been bought in from other factories to complete a specific service commission as highlighted above. We shall refer later to the practice of porcelain manufactories at this time of purchasing waste shards from other sites to supplement their own ground paste mixtures—this component was termed "grog" (Owen and Morrison, 1999). The incorporation of pieces manufactured at other factories into large services to complete their complement is well documented already at Swansea, where the famous *Lysaght* service, decorated locally by Henry Morris (Fig. 2.3), contains several items of Coalport origin porcelain. It is not generally realised that the *Biddulph* service also is in this category, although being London decorated, and several items of Coalport porcelain it is believed also feature in this commission. So, some doubt must be cast upon the definitive attribution of the SW/S9 specimen reported in Table 3.2. As a matter of interest, the analytical compositional data for Coalport porcelain recounted in Eccles and Rackham's work (*Analysed Specimens of English Porcelain*, 1922) give the following figures: SiO_2 42.88%, Al_2O_3 15.06%, H_3PO_4 16.30% and CaO 23.16% (Na_2CO_3 and K_2CO_3 were not determined). Comparison with the Swansea *Biddulph* service specimen analytical data given in Table 3.2 by Owen et al. (1998) shows that it might be difficult to discriminate between Swansea duck-egg porcelain and Coalport porcelain of the same period using these data alone—and it may be concluded that the *Biddulph* specimen, therefore, could well be of Coalport porcelain origin. The main reason for this seems to be that we do not as yet have an idea of intersample compositional range limits which apply to formulations prepared over even a narrow temporal range as there is not sufficient information acquired from the number of replicates required to affect this measurement: so, for example, does a 2% change in silica, a 5% change in alumina, a 3% change in phosphoric acid and a 6% change in lime actually confirm the diagnostic variations as sufficiently significant to categorically and unequivocally determine the origin of the sample as either Coalport or Swansea as individual compositional variations at each factory could bring the perceived discrepancies even closer? One would like to believe that would be the case, but small errors in weighing out the initial components at the factory, for example, could account for a significant error bar in these elemental oxide determinations: a simple error amounting to 0.5 lb in the bone ash aliquot of 26 lbs in manufacture of the putative Nantgarw porcelain paste mixture will give rise to a 2% error in the analytical bone ash content (and hence in the phosphoric acid determination) and a similar error in the 20 lbs aliquot of china clay will produce an error of 3% in the silica analytical determination. Both of these are close to the defining compositional differences between Coalport and Swansea china—similar errors in weighing of the components at each of the two factories at any particular time gives error bars of 4% in bone ash and 6% in silica determinations, respectively! Likewise, without knowledge of the degree of water retention offered by the individual components, which is related to their mine locations, carriage transportation and ultimate storage conditions, then further errors could arise from this source in the weighing out of the paste recipe ingredients at the factory. We do not have any information about the accuracy or otherwise of preparation of the paste recipes originally—and it is here that we are most likely to find that serious quantitative approximations apply: the accuracy of weighing of these components is inestimable—it would be reasonable to assume this would be correct to the nearest pound or fraction of a pound, possibly, and this could even then result in differential compositional variations which will necessarily affect the comparison of the resulting analytical data determinations of the finished product. This would be reflected in the possibly incorrect interpretation of analytical data of specimens as being evidence of an experimental variation being made to the porcelain body—a "tweaking" of component compositions having been effected rather than a simple weighing error on certain components which was deemed to be acceptable in house at that time.

- A more serious analytical question centres upon the methodology used to determine the individual percentages of the components: Eccles and Rackham (*Analysed Specimens of English Porcelain*, 1922) do not state anywhere in their account how they actually analysed their specimens, the number of replicates or elemental determinations used and what methods were adopted for their determination of the elemental oxide components. This would not be acceptable today in any scientific paper as it is essential that the procedures used in the obtaining of the data are given so that the work could be verified independently, if desired. Hence, we have no idea specifically what gravimetric or volumetric methods were used in the determination of silica, alumina and phosphorus—or indeed how many replicates were used to assess the precision of the measurements in the original analytical work form which defined error measurements can then be assessed. Furthermore, although the determination of phosphorus is uniquely related analytically to the bone ash content of the original mixture, a very important stage of the calculation is missing in that we have no knowledge of the chemical formulation adopted by the analyst to deduce the amount of the bone ash component: on examination of their data it can be deduced that for all three specimens analysed by Eccles and Rackham the conversion of phosphoric acid into bone ash content is achieved by the use of a multiplication factor of 2.47—although the origin of this parameter is lost, yet the bone ash content is actually a vital piece of comparative inter-factory and even intra-paste information within a factory! This situation is exacerbated by the discovery in the decade following Eccles and Rackham's work by the revelation that the compositional formulation of bone ash had been improperly and incorrectly formulated hitherto (Morgulis and Janacek 1931) and Eccles and Rackham do not specify the formulation that they have adopted for their bone ash component: hence, the back-calculation of phosphorus content in bone ash from their determined phosphoric acid is potentially seriously compromised. It appears that prior to the definitive work of Morgulis & Janacek, the chemical formulation of bone ash was considered to comprise mainly dibasic calcium phosphate dihydrate, $CaHPO_4.2H_2O$, but Morgulis and Janacek showed that it was actually better described as calcium hydroxyapatite, $[Ca_3(PO_4)_2]_6$. Ca $(OH)_2$, from which two formulae we can estimate the % P as 17.92 and 9.52%, respectively. Clearly, such a difference in phosphorus content in the formulation of one of the major components of porcelain paste before firing will have a significant effect upon the interpretation of the analytical results determined from the specimens used.

- A further point of concern analytically, which renders it difficult to compare the four sets of data under consideration here, namely those of Eccles and Rackham, Tite and Bimson, Owen et al. and Owen and Morrison, is that each set expresses the phosphorus determined in different ways: Eccles and Rackham as phosphoric acid, H_3PO_4, Tite and Bimson and Owen and Morrison as phosphorus pentoxide, P_2O_5, and Owen et al. as the orthophosphate ion, PO_4^{3-}. Each of these have three different component percentages of elemental phosphorus: namely, 31.6, 43.6 and 32.6%, respectively. When converted into relative bone ash content, therefore, several inconsistencies can be expected. Table 3.4 gives a listing of the phosphorus determinations cited in the three reported studies correlated with corrected percentages based on a single P_2O_5 standard for better comparison: the last column in Table 3.4 gives the estimated bone ash content derived from the corrected phosphorus content for analysed Swansea and Nantgarw porcelains. The implication is, therefore, that it is not strictly correct to compare the percentage compositions derived directly from the three independent analytical sets of results for Swansea and Nantgarw porcelains without such a correction having been applied because of the potentially different methods used to deduce the component concentrations in the original paste, although component concentrations in internal specimen comparisons should still be valid for comparison purposes.

The corrected phosphorus percentages shown in Table 3.4 are converted into bone ash figures using the multiplication factor derived from Eccles and Rackham (*Analysed Specimens of English Porcelain*, 1922) the 18 Nantgarw porcelain specimens analysed have a bone ash content ranging between 1% and 47%, with the three specimens analysed by Tite and Bimson from the British Museum excavation of shards giving a close bracketing range of 42 ± 1%. This latter value agrees very well indeed with the recipe formulation for Nantgarw porcelain allegedly revealed to John Taylor by Samuel Walker in 1847 and reproduced by Professor Mellor's experiments in 1885, which calculation gives as 42% bone ash. The other fifteen experimental data values are of a significantly different range compared with the values given by Tite and Bimson. Whereas Eccles and Rackham quote just two values of 35 and 40%, respectively, for bone ash content in their two specimens analysed, Owen et al.'s examination of their three specimens for phosphorus content in samples NG8, NG/N2 and NG/N4

Table 3.4 Analytical data for Bone Ash Content of Swansea and Nantgarw Porcelains Corrected for Different Phosphatic Standards to P_2O_5

Specimen	% Age Phosphorus (cited variously as H_3PO_4, P_2O_5, PO_4^{3-})				
	E and R	T and B	O et al.	O and M	Bone Ash (%)
NG21	10.10				35
NG22	2.40				40
NG14/1		17.4			43
NG18/4		16.4			41
NG18/7		16.4			41
NG8			13.2		33
NG/N2			5.6		14
NG/N4			0.4		1
N11				15.4	38
N12				15	37
N13				17.1	42
N14				17.4	43
N23				19.1	47
N24				17.8	44
N25				0.5	1
N37				16.3	40
N43				0.4	1
N44				0.5	1
SW23	7.14				25
SW40					
SW41	0.16				1
SW/S1					
SW7		0.2			1
SW/S9		10.0			25
GVLE	9.32				30

reveals that the bone ash content of each is anomalously low, being just 33, 14 and 1%, respectively. The later work of Owen and Morrison analysed ten further shards, of which three samples had just 1% bone ash content with an average bone ash content of only 29%. The 13 shard samples which provided the basis for these last two analyses could thus be suspect in that they were described as having been collected from specimens found "*lying around*" the factory site: in a strict forensic evaluation, of course, they cannot be assumed to have originated from the china works at Nantgarw just because they are "lying around" at the site as in the intervening 200 years or so since the factory closure they could have been left as broken or discarded detritus from other events or operations carried out there. There was much activity being carried out at the Nantgarw site after the hiatus caused by the premature death of Thomas Pardoe in 1823, and William Pardoe re-opened the works as a going concern in 1835 for the manufacture of clay pipes until that operation ceased due to the worldwide production and distribution of ready-rolled cigarettes by the American Tobacco Company in the mid-1880s. Hence, an assortment of china waste and detritus could have accumulated at the Nantgarw site up to the closing years of the 19th Century. On a related point, the archives held at the Derby China Works (Anderson 2000) reveal that William Duesbury did purchase quantities of bulk waste fired china shards from other factories to experiment with their incorporation as ground material in his porcelain paste. Owen et al. and Owen and Morrison suggest strongly that there is evidence here for a very low almost non-phosphatic and very highly silicious Nantgarw body which they allege could have resulted from experiments carried out by Billingsley and Walker at Nantgarw in an attempt to improve the robustness of their porcelain, just as Dillwyn and Walker had experimented at Swansea in the previous two years when they created the new trident body in replacement for the duck-egg paste. Owen and Morrison cite ultraviolet radiation transmission colours of different hues from the highly phosphatic *versus* the highly silicious Nantgarw shards as additional support for their thesis and suggest that an examination of finished Nantgarw porcelains in ultraviolet light may answer the question of whether or not there is a second Nantgarw body. Certainly, the analytical data derived from the experiments of Owen and Morrison needs explanation—as to the true identity of their shards: are these genuine examples of a hitherto unsuspected Nantgarw silicious body which analytically is significantly different from the accepted highly phosphatic version or are they interlopers and outliers whose location at the factory site could be considered suspect.

What is clear, despite the conclusions of the analysts, is that these outliers described above cannot unambiguously be considered as forensic evidence of a new Nantgarw body which contained very small and significantly different amounts of bone ash compared with the others analysed, until the location of the Nantgarw waste dump is located and similar shards can be excavated archaeologically in a chronological context which will provide a temporal parameter for evaluation against surviving finished porcelain specimens.

The Swansea porcelain analyses too can be compared in a similar fashion: only four specimens report a bone ash presence and the variation is significant—two each have a composition of 25% and the other two each of only 1%. The first of these, SW23, is from a damaged specimen in the Victoria & Albert Museum, analysed by Eccles and Rackham in 1922 (*Analysed Specimens of English Porcelain*). They comment that the comport dish was impressed Swansea and was simply decorated, probably signifying a locally decorated piece: it is not seen to be an example of the duck-egg porcelain body esteemed by Dillwyn, whose composition was reported above and for which we have calculated a bone ash content of 44%, very similar to that of Nantgarw porcelain as determined by Tite and Bimson. It could be conjectured, therefore, that this is an example of a modification in paste body which decreased the bone ash content and increased that of the china clay—experiments which we know Dillwyn was undertaking in an effort to increase the robustness of his china, albeit eventually achieved at the expense of its translucency. Support for this hypothesis is afforded by the high alumina content of this piece, analysing at over 26%, which can be ascribed to the presence of an increased percentage of china clay aluminosilicates. It appears therefore that the comport dish could be an example of a trident body formulation from this analytical evidence, although this is not commented upon by the analysts. The other two Swansea specimens analysed in the Eccles and Rackham report (*Analysed Specimens of English Porcelain*, 1922) have a zero percentage of bone ash component but correspondingly larger percentages of silica and magnesia, with lower alumina content—all conforming to an increased content of flint glass, lower china clay content and a higher steatitic, magnesium silicate, component giving rise clearly to a more extreme "*trident porcelain*" category assignment. Tite and Bimson did not analyse Swansea porcelain in their study, but Owen et al. reported the analysis of nine Swansea specimens, of which eight are clearly not phosphatic with phosphate percentages between 0.1 and 0.2% only, a low lime content between 0.3 and 0.4%, but all with high silica (70–74%) and a high alumina content (22–25%). The remaining sample, SW/S9, has only a 25% bone ash content but with a lower silica content (45%) and high alumina content (20%) and lime content (17%). The latter specimen is from the named *Biddulph* service and was taken from a decorated plate in the Royal Institution of South Wales at Swansea—despite the lower bone ash content of this piece at 33%, it nevertheless is classified by Owen et al. as an example of duck-egg porcelain. Some doubt has recently been cast on the authenticity of origin of several items in this service as being completely "Swansea"—it could be, therefore, that this is an example of an interloper and further work needs to be done to clarify this. In fact, the analytical figures provided by Owen et al. for this specimen are comparable very closely with the Coalport analyses provided by Eccles and Rackham in their study and perhaps seem to lend credence to this hypothesis. Certainly, as commented upon by Owen et al., this is not a *trident* body, so it remains a mystery—Owen et al. have suggested that it constitutes a new porcelain body type for Swansea and consistent with modifications towards a Nantgarw-type paste introduced by Dillwyn upon the arrival of Billingsley and Walker at Swansea, hitherto unrecognised, but the presence of an associated piece or interloper should not be completely excluded. Again, the importance of the discovery of chronologically secured wasters from the factory dump would be of invaluable assistance here as these would provide a temporal context for the *Biddulph* service items: according to its commissioning, this service would have been prepared during the 1817–1819 period when Swansea duck-egg porcelain was in its ascendancy and not as early as Billingsley and Walker's arrival at Swansea from Nantgarw in 1812 after which Swansea glassy porcelain was being produced.

3.9 Nantgarw and Swansea Porcelain: Minor Additives

A further comment can be made about the presence of trace quantities of chemical components such as iron oxide, titanium dioxide and sulphite which have been identified in the later analyses. The former, noted first by Eccles and Rackham (*Analysed Specimens of English Porcelain*, 1922), can be correlated with the use of sand, where naturally occurring impurities such as iron oxide in the quartz can be found. This is probably the reason for Dillwyn incorporating into his recipe a small concentration of the blue glass smalt, cobalt aluminosilicate, for his duck-egg porcelain paste as the blue colour would negate the effect of the yellow colour that the presence of iron oxides might have upon the reflected and transmitted light. The second of the trace components, titanium dioxide, is a white pigment found in small quantities in kaolin or china clay, with which it occurs naturally—hence the geographical sourcing of different clays or outcrops could provide a variation in concentration of this trace material. Titanium oxide occurs naturally as three mineral polymorphs, anatase, rutile and brookite and rutile is the thermally most stable form at the higher temperatures experienced in ceramic kiln firing, although anatase is found commonly as an impurity in Cornish china clays. Finally, the presence of sulphite ions, SO_3^{2-}, found by Owen et al. in the Swansea shards SW/S1, SW7 and SW/S9 seems to be indicative of a reducing kiln atmosphere, as otherwise an oxidation would have occurred to sulphate ions, SO_4^{2-}. This deserves further consideration as all three Swansea shards exhibit the trace presence of sulphite at the level of a fraction of 1%. It has been proposed by Dr. John (*Nantgarw Porcelain*, 1948) that a possible reason for the characteristic iridescence of London-decorated Nantgarw porcelain is the use of a reducing enamelling kiln in the London ateliers by artists who decorated the porcelain purchased in the white for the London clientele. If so, this could explain the presence of the sulphite on the SW/S9 shard, which belonged to the London-decorated *Biddulph* service, prepared in the atelier of James Bradley and decorated there by Philip Ballard. However, it cannot explain the presence of a similar trace determined in the other two Swansea shards which were fired locally: also, it would be reasonable to suggest that the sulphite would be found at the interface between the applied enamel pigments and the glaze in the London -fired material and we have no knowledge of its precise location from the analytical work carried out.

The presence of other analytes labelled as calcium oxide, magnesia, potash and soda also relate to contributions from several other components, such as phyllosilicates, steatite, china clay and limestone additives. In some cases, it has been proposed that the limestone, calcium carbonate, which thermally decomposes to calcium oxide at a temperature of around 700 °C and helps to create a fusible alkaline flux for the body, could also contain some magnesium from dolomitized limestone. Dolomite itself contains equal proportions of calcium and magnesium carbonates and is formulated as $CaMg(CO_3)_2$, hence it would be converted to calcium and magnesium oxides in kilns used to prepare the initial frit. Limestone is represented by calcium carbonate and magnesite by magnesium carbonate, which are the extremes of the generic dolomitized limestone category of which dolomite contains equimolar amounts of calcium and magnesium carbonates. Lewis Dillwyn makes a comment relating to the overuse of limestone in the frit, cautioning against this because of the tendency for improperly converted calcium carbonate to decompose to lime during the final firing process, during which gas bubbles of carbon dioxide are produced which create defects, so ruining the final porcelain body texture and appearance, as well as its transmission characteristics and translucency. Finally, the presence of trace amounts or very small concentrations of lead in the analyses can be clearly related to the presence of flint glass in the frit used in the paste, which was added in varying amounts at Swansea, particularly. It is equally possible of course that the lead determinations in the analyses arise from a contamination of the ground specimens upon sampling with particles of the applied lead-rich glaze used: this would be particularly relevant for Eccles and Rackham's work (*Analysed Specimens of English Porcelain*, 1922) which involved decorated and glazed china items in finished form, but not so for that of Owen et al. (1998) who used a mixture of glazed and biscuit shards. A relevant analytical comparison can be made in this context of the change in concentrations of the soda and potash % ages, which can be derived from several sources, but an important possible source would have been the substitution of crown or soda glass for lead glass in the initial frit mixture for the paste, which would be revealed in analytical figures which gave increased amounts of sodium and potassium and a decreased lead figure. Other sources of potassium and sodium are polymeric silicate mixtures and clays. This discussion is very relevant for information regarding the sourcing of materials for preparation of the initial frit for Swansea and Nantgarw porcelain manufacture: forensically, modern analytical methods provide a powerful means of potentially accessing source information geologically and geographically from the minor components present in the raw materials such as china clay, soapstone, feldspar and limestone. The sourcing of china clays in particular for porcelain production is highly significant for the resultant behaviour of the paste body composite in the firing kilns as has been mentioned in letters correspondence relating to porcelain manufacture. Dillwyn has commented that the presence of an alabaster onyx marble impurity in the natural mineral materials he used for porcelain production at Swansea resulted in a most unacceptable and undesirable "blistering" effect on the fired porcelain body, which can be ascribed to carbon dioxide gas formation in the incipient porcelain body during firing: calcium carbonate in its several forms, such as limestone, aragonite and marble, decomposes at about 675 °C into calcium oxide and carbon dioxide gas, the latter would be trapped in a dense porcelain silicate matrix being fired and cause bubble blemishes in the final product. The importance of sourcing materials from selected mines was therefore of paramount importance to porcelain manufacturers. Hence, Lewis Dillwyn's insistence on securing the product of the St. Stephen's mine for his normal china clay and the Norden mine, both in Cornwall, for his specialised ball clay. Likewise, William Billingsley would have demonstrated a similar aptitude for securing supplies of his chosen clay and glass frit raw materials ideally from specialised mining operations: John (*Nantgarw Porcelain*, pp. 49–53, 1948) points out that Billingsley and Walker took great pains to keep their recipe for the Nantgarw porcelain body secret and he also alludes to the fact that this was, perhaps surprisingly, also not communicated to their sponsors, William Weston Young at Nantgarw and Lewis Weston Dillwyn at Swansea. However, Dillwyn has left some rather sparse notes about the composition of Swansea pastes he was particularly interested in as we have seen above (Appendix B) and Young has done the same to a much lesser extent for Nantgarw—with the proviso in the latter case that Young's notes address the glaze composition in particular for usage by Thomas Pardoe, so that they were able to complete the decoration of unsold Nantgarw stock held in store locally after Billingsley and Walker's departure from Nantgarw for

Coalport in 1819. Modern observers have noted that the Pardoe glaze is seen to be thinner and rather more creamy in appearance compared with the very white glaze used earlier by Billingsley and Walker. Early experiments designed to reproduce the Nantgarw porcelain body were undertaken by a ceramic research chemist, Professor J. W. Mellor, at Stoke- on- Trent based on this body recipe which were then fired at 1250–1300 °C to yield a porcelain which was considered as good as that of the original Nantgarw when subsequently glazed according to Young's notes (see Nance, *The Pottery and Porcelain of Swansea and Nantgarw*, pp. 389–394, 1942). From this, Dr. John has deduced that fine-grained Lynn sand from Norfolk was mixed with calcined flints from the same source, pearl ash (crude potassium carbonate, soda ash (sodium carbonate), bone ash from calcined ox bones and powdered china clay (possibly with some associated amounts of feldspar and soapstone) were used in the Billingsley/Walker Nantgarw porcelain composition. The composition of the Nantgarw glaze was also given as follows: 50 parts sand, 60 parts borax, 20 parts whiting, 4 parts nitre and 4 parts lead oxide were sifted well and fritted, then to 50 lbs of this frit was added a mixture of 30 parts china stone, 4 parts china clay, 4 parts whiting and 4 parts lead oxide. Professor Mellor made up a glaze according to this recipe and fired it onto the Nantgarw simulated biscuit porcelain made as above at a lower temperature of 1100–1120 °C. In the joint opinion of both Professor Mellor and Mr Nance, the body and glaze of the resultant porcelain very closely represented the complete Nantgarw porcelain produced by William Billingsley in the white. A further recipe for a Nantgarw glaze post –Billingsley/Walker period has also been reproduced by Nance (*The Pottery and Porcelain of Swansea and Nantgarw*, p. 393, 1942), and attributed to William Weston Young and Pardoe for the express purpose of completing and decorating the china left in stock for disposal locally between 1819 and 1822. This information purportedly arose from relevant notes transcribed from Young's diary given to Turner (*The Ceramics of Swansea and Nantgarw*, 1897): this glaze had the composition of 5 parts of Lynn sand, 4 parts borax as a frit then the addition of either 1 or 2 parts of lead oxide to one of the frit. Professor Mellor found that both glaze recipes gave adequate representation of the so-called Pardoe glaze seen on later Nantgarw wares but it was not made clear which of these was believed to have been actually used by Young and Pardoe. Although the actual sourcing of their china clay by Billingsley and Walker has now been lost it is also quite possible that they added small quantities of unknown materials to their recipe: Dillwyn added arsenic oxide, smalt, borax and ball clay to his mixture at Swansea, for example, and such items could well have appeared in the Nantgarw paste mixture too. It would be reasonable to suppose, without any evidence either for or against, that Billingsley and Walker in the second phase of their Nantgarw porcelain production upon their departure from Swansea would have also acquired the best china clay from Cornwall for their efforts, knowing of the success of its incorporation by Dillwyn into Swansea porcelain.

3.10 Summary

In summary, it can be concluded that in over a hundred years of chemical analysis of Welsh porcelains, very little data has been acquired from finished, damaged pieces and none at all from perfect pieces. Hence, Eccles and Rackham (*Analysed Specimens of English Porcelain*, 1922) reported their analyses on fragments of two Nantgarw and three Swansea pieces, all damaged, and Owen reported a solitary analysis on one Swansea piece, also damaged, all of which required destructive analytical procedures of some form or another. Clearly, there is a need for the acquisition of analytical data from perfect pieces of Nantgarw and Swansea porcelain which involves essentially rigorous non-destructive procedures where chemical or mechanical pre-treatment is forbidden. Later in this treatise we shall explore the analytical data emanating from some preliminary Raman spectroscopic analyses on Nantgarw and Swansea porcelains to assess the future development and application of this technique in this area: the initial procedure will be to evaluate the potential of utilising the Raman data for the identification of the individual factories and the discrimination between them. It will be interesting to see if different body paste compositions are reflected in the Raman spectroscopic signals, which will afford a means of assessing factory paste changes during a production history—such as Lewis Dillwyn's change in his Swansea body recipe from glassy to duck-egg and finally to trident porcelains. An integral part of this examination will be the accessing of analytical data relevant to body composition through the applied glaze or enamels on perfect porcelain pieces. Finally, it is possible that the nature of the surface blemishes which created problems for the post-firing acceptance of quality porcelains can be quantified through a micro analysis of the imperfections.

References

J.A. Anderson, Derby Porcelain and the Early English Fine Ceramic Industry. Ph.D. Thesis, University of Leicester, UK, Oct 2000

E.A. Barker, *Pottery and Porcelain in the United States* (New York, 1893), p. 178

D. Battie, The Eyes Have It. *Antique Collecting J. Antique Collectors Club* **28/9**, 3, (1994)

M. Baudoin, Procede pour la Determination des Mineraux Constituent les Haches Prehistorique Metaliques Employ de l'Analyse. Comptes Rendus Hebdomadais de Seances de l'Academie de Sciences **173**, 843–846 (1921)

W.F. Broderick, *An English Porcelain Maker in West Troy*, vol 5(2) (The Hudson Valley Regional Review, 1988), pp. 23–40

W. Burton, *A History and Description of English Porcelain* (Cassell & Co., London, 1902)

W.B. Chaffers, *Marks and Monograms on Pottery and Porcelain with Historical Notes on Each Manufactory* (J. Davy & Sons, London, 1863) (Kessinger Legacy Reprints, Kessinger Publishing, Whitefish Montana USA, 2010)

R.A. Chalmers, F. Szabadvary, Jons Jakob Berzelius (1799–1849) and Analytical Chemistry. Talanta **27**, 1029–1050 (1980)

A.H. Church, *English Porcelain: A Handbook to the China Made in England During the 18th Century as Illustrated by Specimens Chiefly in the National Collection* (A South Kensington Museum Handbook, Chapman & Hall Ltd., London, 1885 and 1894)

P. Colomban, Raman Spectrometry: A Unique Tool to Analyze and Classify Ancient Ceramics and Glasses. Appl. Phys. A **79**, 167–170 (2004)

P. Colomban, A Case Study: Glasses, Glazes and Ceramics, in *Raman Spectroscopy in Archaeology and Art History*, ed. by H.G.M. Edwards, J.M. Chalmers (Royal Society of Chemistry Publishing, Cambridge, 2005)

P. Colomban, The Destructive/Non-destructive Identification of Enameled Pottery, Glass Artifacts and Associated Pigments—a Brief Overview. Arts **2**, 77–110 (2013)

P. Colomban, V. Milande, H. Lucas, On-site Raman Analysis of Medici Porcelains. J. Raman Spectroscopy **35**, 68–72 (2004)

E.E. Davis, The Pottery Notebook of Maude Robinson: A Woman's Contribution to Art, ca. 1903–1931. Ph.D. Thesis, University of Michigan, 2007, Published by ProQuest International Information and Learning Co., Ann Arbor, Michigan, USA, 2007

Sir H. Davy, Some Experiments and Observations on the Colours Used in Paintings by the Ancients. Philos. Trans. R. Soc. **105**, 97–124 (1815)

L.W. Dillwyn, *Notes on the Experimental Production of Swansea Porcelain Bodies and Glazes.* Made by Lewis Weston Dillwyn with Samuel Walker at the Swansea China Works Between 1815 and 1817. Presented to the Library of the Victoria &Albert Museum, South Kensington, London by John Campbell in 1920 (Reproduced in Eccles and Rackham, *Analysed Specimens of English Porcelain*, 1922, see reference)

H. Eccles, B. Rackham, *Analysed Specimens of English Porcelain in the V&A Museum Collection* (Victoria & Albert Museum, South Kensington, London, 1922)

H.G.M. Edwards, *Swansea and Nantgarw Porcelain: A Scientific Reappraisal* (Springer Publishing, Dordrecht, The Netherlands, 2017)

H.G.M. Edwards, P. Colomban, B. Bowden, Raman Spectroscopic Analysis of an English Soft-paste Porcelain Plaque-mounted Table. J. Raman Spectrosc. **35**, 656–661 (2004)

J. Haslem, *The Old Derby China Factory* (George Bell, London, 1876)

M. Hillis, The Development of Welsh Porcelain Bodies, in *Welsh Ceramics in Context, Part II*, ed. by J. Gray (Swansea, Royal Institution of South Wales, 2005), pp. 170–192

W.D. John, *Nantgarw Porcelain* (Ceramic Book Co., Newport, 1948)

W.D. John, *Swansea Porcelain* (Ceramic Book Co., Newport, 1958)

A.E. Jones, L. Sir, *Joseph, Swansea Porcelain: Shapes and Decoration* (D. Brown and Sons Ltd, Cowbridge, 1988)

S. Morgulis, E. Janacek, Studies on the Chemical Composition of Bone Ash. J. Biol. Chem. **93**, 455–466 (1931)

E.M. Nance, *The Pottery and Porcelain of Swansea and Nantgarw* (B.T. Batsford Ltd., London, 1942)

J.V. Owen, R. Barkla, Derby Porcelains: Recipe Changes, Phase Transformations and Melt Fertility. J. Archaeol. Sci. **24**, 127–140 (1997)

J.V. Owen, J.O. Wilstead, R.W. Williams, T.E. Day, A Tale of Two Cities: Compositional Characteristics of Some Nantgarw and Swansea Porcelains and Their Implications for Kiln Wastage. J. Archaeol. Sci. **25**, 359–375 (1998)

J.V. Owen, M.L. Morrison, Sagged Phosphatic Nantgarw Porcelain (ca. 1813–1820): Casualty of Overfiring or a Fertile Paste? Geoarchaeology **14**, 313–332 (1999)

J.V. Owen, A New Classification Scheme for 18th Century American and English Soft Paste Porcelain, in *Ceramics in America*, ed. by R. Hunter (USA, Chipstone Foundation, Milwaukee, Wisconsin, 2002), pp. 45–61

A.M. Pollard, From Bells to Cannon—The Beginnings of Archaeological Chemistry in the Eighteenth Century. Oxf. J. Archaeol. **32**, 335–341 (2013)

A.M. Pollard, Letters from China: A History of the Origins of the Chemical Analysis of Ceramics. AMBIX (Society for the History of Archaeology and Chemistry) **62**, 50–71 (2015)

E.G. Ramsay, W.R.H, Ramsay, in *Bow: Britain's Pioneering Porcelain Manufactory of the 18th Century*. The International Ceramics Fair & Seminar (Park Lane Hotel, London, 2007a), pp. 1–16

W.R.H. Ramsay, E.G. Ramsay, A Classification of Bow Porcelain from First Patent to Closure: c. 1743–1774. Proc. R. Soc. Vic. **119**(1), 1–68 (2007b). ISSN 0035-9211-1-168

W.R.H. Ramsay, E.G. Ramsay, *The Evolution and Compositional Development of English Porcelains from the 16th C to Lund's Bristol c. 1750 and Worcester c. 1752—the Golden Chain* (Invercargill, New Zealand, 2017)

E.E. Richards, Romano-British Mortaria: Analytical Results. Archaeometry **2**, 23–31 (1959)

J. Taylor, *The Complete Practical Potter* (Shelton, Stoke-upon-Trent, 1847)

W. Turner, *The Ceramics of Swansea and Nantgarw* (Bemrose & Sons, Old Bailey, London, 1897)

M.S. Tite, M. Bimson, A Technological Study of English Porcelains. Archaeometry **33**, 3–27 (1991)

J. Twitchett, *Derby Porcelain 1748–1848: An Illustrated Guide* (Antique Collectors Club, Woodbridge, Suffolk, 2002)

A.I. Vogel, *A Textbook of Quantitative Inorganic Analysis: Including Elementary Instrumental Analysis*, 3rd edn. (Longman's Green & Co., Ltd., London, 1961)

Chapter 4
Components of Porcelain Manufacture

Abstract A detailed description of the individual components and minerals utilised in the raw materials of porcelain manufacture such as sand, smalt, borax, alum, ball clay, kaolin, soaprock, steatite, pearl ash, alabaster, flints and flint glass is followed by a consideration of the influence of the high temperatures in the kilns and their conversion into different mineral components. For example, the creation of new minerals such as wollastonite, whitlockite, mullite and sanidine in porcelains fired to temperatures in excess of 1300 °C, typical of the Swansea and Nantgarw kilns.

Keywords Raw materials · Mineral conversion · High temperature firing Sulfur · Flint glass · Steatite · Soaprock

It is appropriate at this point to consider the individual mineral components of porcelain manufacture as these will dictate the durability and eventual translucency of the finished article after firing at high temperatures.

Sand: the finest quartz river sand was specified by many business directors to provide the silica, SiO_2, necessary for preparation of the basic porcelain paste fusion matrix. Sand comprises generically finely-divided rock and mineral particles, geologically occupying the particulate range between gravel and silt. Fine river sand, or white quartz sand, has a particle size range broadly between 0.06 and 0.1 mm and is often coloured by the presence of transition metal oxides, such as iron (III) oxide, Fe_2O_3, and infusible particulate matter such as carbon or manganese (IV) oxide, pyrolusite, MnO_2. Although a pure white sand is found which can also contain gypsum, sand derived from the weathering of basalts can display a range of colours from green to brown to black, due to the presence of minerals such as olivine and magnetite and these can generate blemishes in the glassy matrix formed upon fusion in the firing kilns, with a resulting deleterious effect on the porcelain translucency and in some cases rendering the final article unsaleable. When occurring near the surface, however created, these blemishes often cause a pitting or grittiness in texture which is not acceptable: in several cases, therefore, ceramic artists have disguised such

H. G. M. Edwards, *Nantgarw and Swansea Porcelains*,
https://doi.org/10.1007/978-3-319-77631-6_4

surface blemishes in an otherwise fine piece of porcelain by applying an enamelled insect such as a moth or a butterfly, or perhaps a single flower bud, which masks the small area affected—examples of this can be seen on the underside of the rim of a very large Nantgarw platter from the *Farnley Hall* service (Fig. 4.1) which has a trio of mosquito-like insects masking a surface blemish (Fig. 4.2). This prestigious service was commissioned by William Ramsden Hawkesworth Fawkes, an influential London businessman, Yorkshire landowner and MP, probably from John Mortlock's in Oxford Street, London, and would have been decorated in the Robins and Randall atelier: clearly, the acceptance of such a large and otherwise perfect piece of Nantgarw sent out from the factory was not problematic for the purchasers "in the white" for subsequent decoration and the addition of little additional insect motifs on a defective surface area not immediately visible from above was deemed to be normal. It is realised therefore that porcelain manufacturers such as Dillwyn and Billingsley would prioritise their selection of a superior quality of fine sand as being highly important for the realisation of their target of producing a very high-quality porcelain: in this respect, the sourcing by William Billingsley of his Lynn sand from the Norfolk gravel pits can be appreciated.

Smalt: a cobalt blue glass (not to be confused with the mineral smaltite, a cobalt arsenide) which is used finely ground in the glassy matrix to combat or offset any background residual yellow or brown colouration which may occur from impurities in the matrix, especially iron oxides from sand. It is also a component of the glaze slip into which the fired porcelain piece is dipped for a sealing coat of a high gloss, transparent white finish. Cobalt blue glass consists of cobalt (III) oxide, Co_2O_3, added

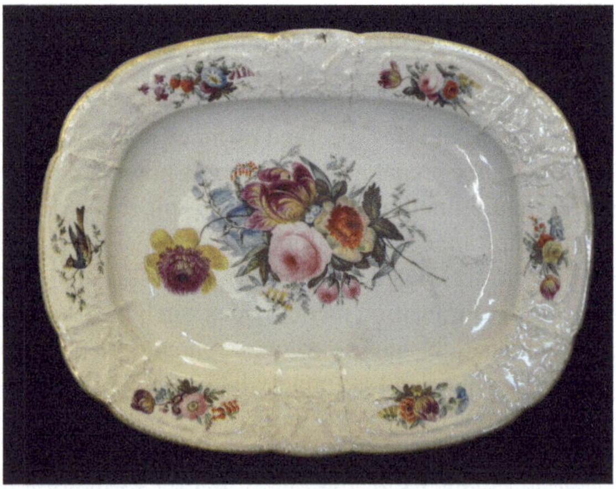

Fig. 4.1 A Nantgarw large porcelain meat platter from the Farnley Hall service, commissioned by William Ramsden Hawksworth Fawkes ca. 1817–1819, showing the superb floral groups on the front surface. London decorated with floral groups and birds in a Brace service arrangement, probably in the workshops of Robins & Randall. Impressed mark NANT-GARW C.W. Reproduced courtesy of Guy Fawkes Esq., Farnley Hall, North Yorkshire

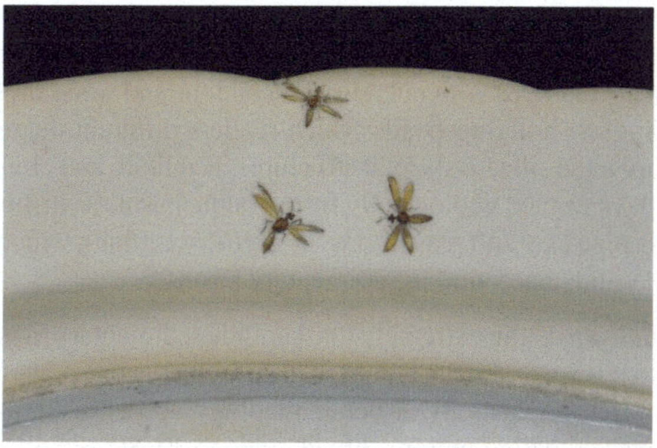

Fig. 4.2 The underside of the rim of the large Nantgarw porcelain meat platter shown in Fig. 4.1, where several mosquito-like insects have been painted to mask blemishes in the porcelain body caused by small black particles. Reproduced courtesy of Guy Fawkes Esq., Farnley Hall, North Yorkshire

to an aluminosilicate glass, typically comprising SiO_2 65%, K_2O 15%, Al_2O_3 5% and Co_2O_3 10%. Frequently represented incorrectly as $CoO.SiO_2$, and also simply as a cobalt aluminate, cobalt blue is really a cobalt aluminosilicate and has been known from about 1500 AD. It has often been confused with the more ancient historic Egyptian blue, which is a calcium copper(II) silicate, $CaCuSi_4O_{10}$, also termed *cuprorivaite*. Egyptian blue was synthesised in antiquity, it is believed around 3500 BC. Cobalt blue was used extensively in blue transfer earthenware patterns in the 18th and 19th Centuries and the famous "*Bristol blue*" coloured glass of bottles and drinking vessels of the Georgian and Regency periods reflect the trade in this valuable coloured oxide through the port of Bristol. Dillwyn recorded his use of smalt in his finest duck-egg Swansea porcelain but is not known if Billingsley adopted this also in Nantgarw: it is suspected that he did not as smalt did not feature in the recipe details provided to Taylor (*The Complete Potter*, 1847), it is believed some years later by Samuel Walker.

Borax: is sodium tetraborate decahydrate, $Na_2B_4O_7.10H_2O$, and is used as a flux to aid the fusion of the glassy components of the paste in the firing kiln—this meant that the paste would become more fusible at a lower temperature and this would assist in the better preservation of the articles in the kiln.

Alum: alum is a hydrated aluminium potassium sulfate, $KAl(SO_4)_2.12H_2O$, and has been described as an unusual additive to porcelain bodies (Ramsay and Ramsay 2007b). Church (1881) has stated that the paste of *Vieux Sevres*, also known as *pate tendre*, comprises 8 parts marl, 17 parts chalk, and 75 parts glass frit—the latter component containing 3. Parts per 100 of alum. Ramsay and Ramsay (2007a) draw attention to this unusual additive and wonder where the idea originated, but they also note that early Bow porcelains have been found to contain alum as a minor component. Clearly, the origin of the sulfur presence in derived analytical data can be ascribed variously to gypsum, anhydrite, pyrites or alum and the presence of sulfur is not in itself an indicator of an alum additive.

Clay: clay is a generic name for the geologically weathered materials from feldspar, granites and basalts. It covers many types of structure based on three-dimensional silicate matrices with interstitial metal ions and water molecules, several of which have Si-O-Si bridging bonds which renders different degrees of pliability and hardness upon the silicate skeletal structures. Kaolinite and china clay are primary or residual clays formed geologically from granite basins as distinct from ball clay (also known as pipe clay and plastic clay, or *argile plastique*), which is a secondary or transported detrital clay found in sedimentary basins.

Ball clay: is an extremely fine kaolinic sedimentary material, low in haematite content so it is very white in colour, whose inclusion in a porcelain paste before firing renders a greater plasticity, a better rheology and ease of working. It has a composition approximating within a broad range to 20–80% kaolinite, 10–25% mica and 6–65% quartz and may also contain carbonaceous material such as lignite, which is burned off upon calcination. It derives its name from mining operations which provide this clay in cubes weighing between 15–17 kg with the corners rounded off. It is found in Eocene deposits (40–50 Mya) in SW England and especially in Cornwall and Devon. It is believed that this comprises the "virgin earth" mentioned by Eccles and Rackham (1922) and Church (1894): expected contaminants are siderite, pyrites, anatase, gypsum and dolomite.

Kaolin: is china clay, sometimes called kaolinite, formulated as $Al_2Si_2O_5(OH)_4$, a layered aluminosilicate of the phyllosilicate type and a member of the serpentine group, structurally containing tetrahedral sheets of SiO_4 units linked to sheets of AlO_6 octahedra produced by the weathering of feldspar. The name *kaolin* is derived from a corruption of the Chinese Kauling (alt. Gaoling), the name of a hill near Jauchau Fu, where the material was first mined to produce hard paste porcelain (Dana, *Textbook of Mineralogy*, 1955) near to the first kilns set up at Jingdzehen in Jiangxi province. It is sometimes represented as $Al_2O_3.2SiO_2.2H_2O$, which can assist in the interpretation of the wet chemical analyses of porcelain through the estimation of the silica and alumina contents. It is appropriate here to describe the complex changes which occur in the structure of kaolin when subjected to the heat of a ceramics firing kiln: firstly, dehydration (loss of molecular water) commences at 550–600 °C to form *metakaolinite* with loss of further hydroxyl (OH) groups up to 900 °C. At this temperature, the first chemical skeletal structural change occurs in the silicate matrix, when the original $Al_2Si_2O_5(OH)_4$ undergoes transformation to $Al_2Si_2O_7$ with the formal loss of two molecules of water. At the slightly higher temperature range of 925–950 °C a spinel is formed when two molecules of the dehydrated aluminosilicate form $Si_3Al_4O_{12}$, with the elimination of a molecule of SiO_2. Around 1050 °C three molecules of this spinel now react to form mullite ($3Al_2O_3.2SiO_2$) with the elimination of 5 molecules of silica as SiO_2 in its high temperature form, *cristobalite*. Finally, at the very highest kiln temperatures of 1400 °C, stability is achieved with some internal rearrangements being made to the needle-like structure of mullite. Ramsay and Ramsay (2002) have defined five different types of clay from various sources used at Bow for porcelain manufacture with the following compositional ranges: silica 44.8–62.69%, alumina 0.49–38.4%, magnesia

0.03–29.93% and lime 0.06–0.3%—all of which contain haematite at about 1%. In addition, several recipes at Bow, at Swansea and Nantgarw and at other factories involve a mixture of kaolin and ball clay. obtained from different mines: the typical composition of kaolin is given as silica 44.8%, alumina 38.4%, magnesia 0.03%, potash 3.33 and soda 0.06%, whereas similar values for the blue ball clay are 54, 30.0, 0.51, 3.1 and 0.46%, respectively. Unsubscribed mixtures of these two clays will therefore defer the analytical interpretations even further to that expected for compositional differences between the alternative mine sources and reinforces the earlier point made about the source-dependent variation in composition of the raw materials.

Soaprock or *steatite*: a metamorphic schist silicate which is rich in magnesium—a variant of talc, often formulated as $3MgO.4SiO_2.H_2O$ or $H_2Mg_3(SiO_3)_4$, it is sometimes described simply as an acid magnesium silicate. A talc formed from the metamorphism of ultramafic protoliths such as serpentinite, a typical steatite is comprised of 63.4% SiO_2, 31.9% MgO and 4.7% H_2O), with the minor occurrence of CaO and Al_2O_3 (Anthony et al. 1995). This material has often been identified and correlated directly and incorrectly as it transpires with the *petuntse* (sometimes referred to as *porcelain stone*) of the Chinese hard paste porcelains where it confers a robustness upon the fired porcelain body; it is frequently found in admixture with kaolin but the texture and translucency can potentially be compromised. Nevertheless, Dillwyn used such a mixture in his later attempts to improve the strength of Swansea china, but the resultant "*trident*" paste was not appreciated by clients, falling short of the fine quality of the duck-egg paste it replaced despite the excellent floral art displayed in the enamelling of these pieces. Soapstone generally contains about 1/3rd of its weight of magnesia, MgO, so analysts have multiplied the magnesia value by 3X to obtain the soapstone content of a particular porcelain item. The soapstone content of some porcelains is significant, for example, a Worcester body formulated by Martin Barr in 1800 gives the following recipe—Lynn sand 300 parts and flint glass 15 parts are calcined in a biscuit oven; then of this frit, 300 parts are mixed with 240 parts soapstone for the final porcelain body. This, therefore, represents a soapstone composition of 48%—yet analytical figures on pieces from this same period give a magnesia content of 10.5%, which on the basis of the above conversion factor, represents a soapstone content of only 31.5%. Clearly, this discrepancy gives cause for concern, as there is a 30% error between the determined analytical composition and the established recipe from the factory work books which requires explanation and we need to examine more closely the incipient errors made in the weighing practices of each component, variance in sourcing of the raw materials and mechanical deviations in the frit preparation. It has already been stated that the corresponding Swansea "*trident*" body formulation accepted finally by Dillwyn after his experiments comprises: "*Body No. 2, silica 4 parts, soapstone 1 part and potash ½ part*", so the precision of the weighing and preparation of the paste mixture is seemingly rather vague and open to incipient errors in formulation although on the surface it appears that the trident body soapstone content is only about 19%, considerably less than the Worcester version of Barr, Flight & Barr cited above. Despite this, Dillwyn's departure from a high kaolin soft paste porcelain which proved to be so successful, commercially marketed and appreciated as his famous duck-egg formulation, into the less well appreciated but more robust trident variation containing soapstone proved a commercial disaster, yet the Worcester factory was still able to maintain a successful market edge in this area, despite the significantly higher levels of soapstone used in the porcelain paste. At this point it is relevant to indicate a potential source of confusion which can arise from the rather incautious usage of the three terms *soapstone*, *china stone* and *soaprock*: strictly, these are different mineralogically. *China stone* is a rare medium-grained, feldspar-rich partially kaolinized derivative of granite which is characterised by the absence of iron-bearing minerals and contains quartz, mica, feldspar, kaolinite and traces of fluorspar. This mineral occurs only the neighbourhood of St. Austell in Cornwall in the UK and should not be identified loosely with *petuntse* (*baidunzi*), the Chinese kaolin or porcelain stone, which is also a micaceous and feldspathic material used in the production of Chinese hard paste porcelain. *Soaprock* is different again in that it is a purer form of steatite, a talc schist and chemically described as a hydrated magnesium silicate. These terms are frequently confused in the literature, although the chemically trained Lewis Dillwyn does identify them separately and correctly in his recipes and formulations as FO (representing *china stone*, or alternatively *composition*) and SR (representing *soaprock*). *Petuntse* on the other hand is often mis-described as *china stone* because of its Chinese origins and is more correctly described as porcelain stone as it is of different mineralogy to *soaprock*, *china stone* and *soapstone*. Hence, the former (*petuntse*) can be reserved exclusively for the Chinese hard paste porcelains whereas the latter variants can be regarded as components of English and Welsh soft paste and hard paste porcelains.

Pearl ash: this component has caused some confusion historically in attempts to describe its chemical composition. It is technically potassium carbonate, K_2CO_3, and is usually sourced naturally in carbonaceous deposits along with its calcium analogues—however, it is mined as an impure material and since the isolation of pure potassium carbonate in industrial quantities is rather expensive and quite prohibitive for commercial porcelain manufacture, it was used as a less pure yet refined form of "potash", which gave rise to its estimated presence in analytical determinations simply as K_2O. It was normally baked in a kiln to remove volatile impurities before use.

Alabaster: although chemically this is a translucent form of hydrated calcium sulfate, gypsum, $CaSO_4.2H_2O$, in earlier times this terminology was also applied to translucent forms of calcium carbonate called *onyx marble* (where it is more correctly termed geologically as calcite or aragonite, two forms of naturally occurring calcium carbonate, $CaCO_3$); it is not therefore to be confused with onyx, which is a coloured silicate related to jadeite, a metasilicate of sodium and aluminium formulated as $NaAl(SiO_3)_2$ or alternatively $Na_2O.Al_2O_3.4SiO_2$. Although Josiah Wedgwood was an adherent of the use of gypsum and alabaster in the manufacture of his creamy Queen's ware from 1759, Lewis Dillwyn is on record as saying that the presence of alabaster could be detrimental to the Swansea porcelain body because of its "blistering" effect—this could be ascribed chemically to its decomposition around 650–700 °C, releasing gaseous carbon dioxide and forming lime, calcium oxide, CaO. The lime would not in itself be a problem as this was often added as a constituent in the porcelain paste mix before firing anyway and would be expected to react with the acid silicates in the mixture to form calcium silicates, but the formation of CO_2 bubbles in the paste reactants in the kiln at elevated temperatures would certainly have created voids in the plastic body and give rise to a lumpiness and blistering in the viscous paste. It can be assumed, therefore, that Dillwyn was referring to the calcite or aragonite connotation for alabaster as gypsum would not decompose in the same way in the kiln, merely dehydrating to anhydrite, $CaSO_4$, and thereafter remaining stable. In the latter instance, water molecules are thermally removed from the gypsum at a temperature between 170 and 250 °C; it is reasonable to assume that if Dillwyn was in fact referring to the gypsum connotation for his reference to alabaster then the removal of water at this low temperature would not be viewed as having such a detrimental effect upon the paste undergoing firing as this would still have been moist anyway through the retention of some water at this temperature. Hence, we can infer that Dillwyn was warning about the effect of carbonate decomposition at more elevated temperatures, when the effect of carbon dioxide emission upon the then drier porcelain body would have been more unsustainable in the kiln firing process.

Flint and flint glass: these are chemically similar in that the major component is silica, SiO_2, and indeed historically they both derive from the same root in that flint nodules from South East England were first used as a source of high purity silica by George Ravenscroft in the 1670s to make his renowned heavy potash lead glass of high refractivity and dispersion, which contained nominally 20–60% lead oxide. Whereas flint occurs as glassy nodules in chalk deposits and can be coloured grey to brownish red by metal oxide impurities, especially iron III) compounds, it is relatively easy to select manually the whitish -blue and grey nodules which represent a very high purity silica content with minimal impurities present. Flint is actually a higher quality form of *chert*, which is a fine grained sedimentary rock containing microcrystalline silica occurring in limestone, chalk and marls: other special forms of chert are chalcedony, jadeite and agates. The main difference, therefore, between flint and flint glass is the significant lead content of the latter, which can be detected by chemical analysis, bearing in mind that lead signals can also be observed from the superficial glaze applied to finished porcelains. Essentially, therefore, we must recognise the chemical distinction which appears in recipes which indicate the use of ground frits containing flints and/or flint glass in porcelain pastes. Reference to Lewis Dillwyn's recipes for his Swansea duck-egg porcelain reveals that these involved the addition of flint glass to his frit, whereas the recipe purporting to be from Nantgarw given to Taylor in 1847 (*The Complete Potter*, 1847) by Samuel Walker specifically excludes this additive. Hence, a very positive analytical discrimination between Swansea and Nantgarw porcelains would be expected to be the analytical marker for lead in the former which would be absent in the latter. Unfortunately, this seems to have escaped attention until now and none of the analytical work carried out to date mentions the importance of the determination of lead content in Swansea or Nantgarw porcelains: the silica content determinations do not assist here since both flints and flint glass are high silica content materials and, therefore, both Swansea and Nantgarw porcelains have a major component presence attributable directly to silica.

4.1 The Swansea and Nantgarw Porcelain Bodies and Glazes

A key component of the porcelain bodies of Swansea and Nantgarw is the china clay, which chemically is composed of phyllosilicates formed from the metamorphic geological degradation of basalts, granites and feldspars and the fineness and purity of the china clays defines the porcelain quality after firing (Dana, *Textbook of Mineralogy*, 1955). Generally, phyllosilicate chemistry is complex and involves the coordination of metal ions to silicon-oxygen bonds classified into two types: namely, dioctahedral and trioctahedral. The former has two thirds of the sites in the octahedral layer of skeletal silicon-oxygen atoms occupied by $Al^{3+}$and Fe^{3+} ions whereas the latter has all three sites in the skeletal silicon-oxygen structure occupied by Mg^{2+} and Fe^{2+}, and other divalent cations. For example, kaolinite, $Al_2Si_2O_5(OH)_4$, has the Al^{3+} metal cations occupying octahedral coordination between two-dimensional sheets of Si_2O_5 units. Another dioctahedral phyllosilicate is serpentine, which has Mg^{2+} and Fe^{2+} ions, and kaolinite and serpentine can exist in solid solution admixtures as natural minerals. The hydroxyl groups usually occur coordinated to the metal ions but water molecules also occur interstitially between the sheets of silicon-oxygen atoms, for example, in vermiculite. The thermal chemistry of silica at high temperatures found operating in porcelain firing kilns is complex, but when calcined to a temperature near 1400 °C, the final form of silica is a *beta*-cristobalite called mullite, which is an atactosilicate matrix formed of stable SiO_4 tetrahedra which share neighbouring oxygen atoms in a three dimensional -O-Si-O- lattice. It is this structure which conveys the superior hardness of the final porcelain paste found after firing in the kiln. Although the porcelain body is not porous to water unlike earthenwares, nevertheless a protective glaze is applied usually to the biscuit body for useful wares and this contains a significant lead content which may already supplement that found in the body if flint glass frit has been used in its manufacture. Flint glass comprises silica to which lead oxides have been added in the form of trilead tetroxide, Pb_3O_4, white lead, $2PbCO_3.Pb(OH)_2$, or litharge, PbO—the percentage of lead in flint glass varies greatly from as little as 4% to as high as 60%, thereby increasing the refractive index to 1.7–1.8 compared with ordinary crown glass or soda glass, which has a refractive index of 1.5. It is for this reason that it is impossible to present a chemical formula of flint glass as it is for other components of the porcelain paste. A lead glaze will also contain lead oxide additives or possibly even a supplementation of finely powdered flint glass mixed with china clay in the glazing slip. It appears that four types of glaze can be identified on ceramics: feldspathic, lead, tin and salt. Normally, the lead glaze is found with high quality porcelains and the others are usually associated with majolica, faience and earthenwares as it seems that feldspar can cause translucency problems with clays unless fired to higher temperatures than those usually adopted in glost kilns.

The china clays generically comprise mixtures of phyllosilicates such as mica, talc and pyrophyllite as well as kaolinite and serpentine: being natural minerals they are also found associated with other minerals such as anatase and rutile, the common naturally occurring titanium (IV) oxides. The other, rather rare, naturally occurring titanium (IV) oxide is brookite. This means that the sourcing of china clays for porcelain production is highly significant for the resultant behaviour of the paste body composite in the firing kilns. Dillwyn has commented that the presence of a presumed alabaster onyx marble impurity in the natural mineral materials he used for porcelain production at Swansea resulted in a most unacceptable and undesirable "*blistering*" effect on the fired porcelain body, which we have ascribed earlier to carbon dioxide formation in the incipient porcelain body. The importance of sourcing materials from selected mines was therefore of paramount importance to porcelain manufacturers. Hence, Dillwyn's insistence on securing the product of the St Stephen's mine for his normal china clay and the Norden mine, both in Cornwall, for his specialised ball clay. Whereas kaolin is considered a primary clay sedimentary deposit formed in situ by the breakdown of granites, ball clay (or blue clay) is a secondary deposit which has been formed from further erosion and aqueous deposition and transportation from its primary source: mined as blocks of approximately 25 cm side, ball clay was rounded at the edges and corners for ease of handling—giving its distinctive name. Ball clay was highly plastic on fusion and was frequently mixed with the less plastic kaolin to improve its workability. Both types of clay were found in localised deposits in Cornwall, near St Austell and provided much of the requirements for the English porcelain industry as well as for the Nantgarw and Swansea factories. Likewise, Billingsley would have demonstrated a similar aptitude for securing supplies of his chosen clay and glass frit raw materials: John (*Nantgarw Porcelain*, pp. 49–53, 1948) points out that Billingsley and Walker took great pains to keep their recipe for the Nantgarw porcelain body secret during their lifetime and he also alludes to the fact that this was perhaps surprisingly also not communicated to their sponsors, William Weston Young at Nantgarw and Lewis Weston Dillwyn at Swansea. However, Dillwyn has left some rather sparse notes about the composition of Swansea pastes he was particularly interested in as we have seen above (these are reproduced in Appendix B, in which the shorthand notation of Dillwyn has been simplified and translated into a more readable form) and Young has done the same to a much lesser extent for Nantgarw—with the proviso that in the latter case Young's notes address predominantly the glaze composition in particular for Thomas Pardoe so that they were able to complete the decoration of unsold Nantgarw stock locally after Billingsley and Walker's departure in 1819. Taylor, writing in 1847 (*The Complete Practical Potter*, Shelton, 1847) ascribes the following recipe to Samuel Walker at Nantgarw and given to him before Walker's departure for the USA to set up a ceramics business there:

> 26 lbs bone, 14 lbs of Lynn sand and 2lbs potash were mixed with water then made into bricks and fired in a biscuit kiln. Then the 42 lbs of cooled frit was ground with 20 lbs of china clay and made into the paste.

An interesting anonymous article appeared some years later in the *Pottery Gazette*, December 1st, 1885, which repeated this composition for the Nantgarw porcelain body and additionally described the glaze which was applied. Early experiments designed to reproduce the Nantgarw porcelain body using this recipe were undertaken by a ceramic research chemist, Professor J. W. Mellor, at Stoke-on-Trent based on this information and then fired at 1250–1300 °C to yield a porcelain which was considered as good as that of the original Nantgarw when subsequently glazed according to Young's notes (see Morton Nance, *The Pottery and Porcelain of Swansea and Nantgarw*, pp. 389–394, 1942). From this, Dr John has deduced that fine-grained Lynn sand from Norfolk was mixed with calcined flints from the same source, pearl ash (crude potassium carbonate), soda ash (sodium carbonate), bone ash from calcined ox bones and powdered china clay (possibly with some associated amounts of mineral feldspar and soapstone) were used in the Billingsley/Walker Nantgarw porcelain composition. The composition of the Nantgarw glaze was also given as follows:

> 50 parts sand, 60 parts borax, 20 parts whiting, 4 parts nitre and 4 parts lead oxide were sifted well and fritted, then to 50 lbs of this frit was added a mixture of 30 parts china stone, 4 parts china clay, 4 parts whiting and 4 parts lead oxide.

Professor Mellor made up a glaze according to this recipe and fired it onto the Nantgarw simulated biscuit porcelain made according to the previous recipe at a temperature of 1100–1120 °C. In the joint opinion of both Professor Mellor and Mr Nance, the body and glaze very closely represented visually the complete finished Nantgarw porcelain produced by William Billingsley. The term "whiting" is interesting in that it usually refers either to calcium carbonate or quicklime, but since the former is converted into the latter by thermal decomposition around 650 °C the distinction between the two for porcelain paste fired in the kiln is quite academic; indeed, Dillwyn has mentioned that the presence of calcium carbonate or limestone in the paste mixture can be deleterious to porcelain translucency because of the bubbles of carbon dioxide formed and trapped in the viscous paste due to its thermal decomposition into calcium oxide.

A further recipe for a Nantgarw glaze used in the post–Billingsley/Walker period has also been reproduced by Morton Nance (*The Pottery and Porcelain of Swansea and Nantgarw*, p. 393, 1942), and attributed to William Weston Young and Thomas Pardoe for the express purpose of completing and decorating the china left in stock for disposal locally and at auction between 1819 and 1822. This information purportedly arose from relevant notes transcribed from Young's diary given to Turner (*The Ceramics of Swansea and Nantgarw*, 1897): this glaze had the composition (Morton Nance, *The Pottery and Porcelain of Swansea and Nantgarw*, pp. 393–394, 1942), of:

> 5 parts of Lynn sand added to 4 parts of borax, to which were added either 1 or 2 parts of lead and glass to 1 of the frit.

Professor Mellor found that both glaze recipes gave adequate representation of the so-called Pardoe glaze seen on later Nantgarw wares but it was not made clear which of these was believed to have been actually used by Young and Pardoe. The high percentage of borax added to the glaze recipes of Billingsley/Walker and of Young/Pardoe should be noted in passing: it is believed this resulted in a more easily fusible glaze slip, and both glaze recipes appeared to give a successful result when fired onto the biscuit simulated Nantgarw porcelain prepared by Professor Mellor. An interesting comment has been made by Morton Nance upon his inspection of the reproduced Nantgarw porcelain in that the visual "pigskin" effect of the new glaze upon articles decorated in the Young/ Pardoe period is not reproduced in Professor Mellor's experiments—this he has attributed to the use of better quality modern flint glass in the substitute glaze.

A very important piece of information which is absent from the Nantgarw recipes as taken from these later publications is the source of the china clay—as we have seen from Dillwyn's notes of his Swansea experiments (Appendix B), the specification of the source of the china clay was vital. Unfortunately, the actual sourcing of their china clay by Billingsley and Walker has now been lost. It is also quite possible that Billingsley and Walker added small quantities of unknown material to their recipe: Dillwyn added arsenic oxide, smalt and ball clay to his mixture, for example, and such items could well have appeared in the Nantgarw paste mixture too. It would be reasonable to suppose, without any evidence either for or against, that Billingsley and Walker in the second phase of their Nantgarw porcelain production upon their departure form Swansea would have also acquired the best china clay from Cornwall for their efforts, knowing of the success of its incorporation into Dillwyn's duck-egg and highly translucent Swansea porcelain body.

In 1991, Tite and Bimson carried out the first modern microanalytical study of Nantgarw porcelain using scanning electron microscopy and energy dispersive X-ray diffraction techniques; however, thin sections cut from factory wasters were necessary to undertake these analyses. Three specimens of Nantgarw porcelains from the British Museum Research Laboratories were used, derived from unspecified excavations carried out at the factory site in 1932. The details are summarised in Table 4.1.

The Nantgarw average glaze analysis of the same three specimens gave 62% SiO_2, 12% PbO, 12% Al_2O_3, 1% Na_2O, 2% K_2O, 10% CaO, 1% P_2O_5, and less than 0.5% FeO and MgO. In the specimen N18-7, special note was made of the presence of the mineral phases whitlockite and anorthite in the Nantgarw body.

Generally, *glassy porcelains* are characterised by a lime-rich body (containing between 19 and 27% CaO), unreacted quartz (from the Lynn sand) and wollastonite in a glassy matrix with a lead oxide content of up to 20% reflecting the use of flint glass frit. The lead content can be very variable because of the range of frit compositions used in the paste, typically varying between 10–35%. On the other hand, *bone ash porcelain*, typical of Nantgarw, will contain unreacted quartz and tricalcium phosphate (whitlockite) in a glassy matrix containing anorthite ($CaAl_2Si_2O_8$), which is not found in glassy porcelains. Typically, a bone ash porcelain will comprise 4 parts bone ash, 4 parts Lynn sand, 0.25 part Dorset blue ball clay and 0.25 part gypsum or alabaster. The bone ash therefore represents about 40–45% of the paste mixture.

Table 4.1 Analytical data for Swansea porcelain (data expressed in %)

Analyst	Specimen	SiO_2	Al_2O_3	P_2O_5[b]	CaO	MgO	Fe_2O_3	K_2CO_3	Na_2CO_3	SO_4^{2-}	TiO_2	Equiv. BA/%	SS%
Eccles (1914)	GVLE	44.3	21	12.9	18.1	0.7	1.9	1.0				33	
[c]Eccles and Rackham (1922)	SW23	47.8	26.5	9.9	13.3		3.1	0.2				25	
	SW40	84.0	8.7		1.0	2.5	2.9	0.6					7.5
	SW41	81.6	8.9	0.3	0.7	4.3	3.3	0.6					1.3
Owen et al. (1998)	S1	68.0	8.2		9.7	0.2	0.1	2.5	10.9	0.2			
	S2	71.1	24.9	0.2	0.3	0.2	0.7	1.6	0.4	0.2	0.4		
	S3	73.8	21.6	0.2	0.3	0.2	0.8	1.6	0.6	0.4	0.4		
	S4	70.7	24.5	0.1	0.4	0.2	0.7	1.8	0.7	0.3	0.4		
	S5	70.5	25.1	0.2	0.3	0.2	0.7	1.7	0.5	0.3	0.4		
	S6	71.7	23.9	0.2	0.2	0.2	0.6	1.9	0.6	0.3	0.3		
	S7	73.6	22.1	0.3	0.3	0.2	0.6	1.7	0.6	0.3	0.4		
	S8	70.0	25.4	0.2	0.4	0.2	0.7	1.7	0.5	0.4	0.5		
	S9[a]	45.2	20.0	13	16.7	0.4	0.3	2.8	1.0	0.1			

[a]*Biddulph service*, [b]*Expressed variously as H3PO4 (E&R), P2O5 (T&B), PO43- (O et al.)*,[c]*GVLE: sample from loan exhibition glynn vivian art gallery*

In Nantgarw porcelain it is thought that china clay and china stone were used in place of ball clay from oblique references made in Dillwyn's notebook (reproduced here in Appendix B; Eccles and Rackham, *Analysed Specimens of English Porcelain*, 1922). Hence, a typical Nantgarw paste mixture would comprise 9 parts china stone (feldspar, 12 parts china clay (kaolinite), 12 parts bone ash and 3 parts lime (Hillis, *Welsh Ceramics in Context II*, 2005). There seems to be some conflict here with the claims of contemporary recipes that no additional lime was added, otherwise the $CaO:P_2O_5$ ratio for the paste would be raised from the value expected for hydroxyapatite, which in the firing process converts to calcium orthophosphate, but the situation is complicated further since experiments have shown that at temperatures between 1000 and 1200 °C there is a calcium depletion from the hydroxyapatite and phosphate components and their incorporation into the calcium-rich glassy matrix takes place. Tite and Bimson (1991) have shown that Nantgarw porcelain does in fact contain unreacted quartz, whitlockite and anorthite in the glassy matrix, which confirm its *bone ash* classification.

4.2 The Materials Present in Fired Nantgarw and Swansea Porcelains

Further to the above discussion, it is relevant now to recount the materials which are likely to be present in Nantgarw and Swansea porcelains, an account of which will prepare the way for their eventual analytical characterisation and which will be informative in the analytical protocol designed to differentiate and discriminate between the products of these two factories. This consideration is critical since it is clear that the paste components used in the original recipes, where these are described or available in the records, will not be useful to describe the composition of the final fired porcelain because of the high temperatures to which the mixtures have been exposed during the firing processes and the consequent chemical reactions undergone thermally. Typical kiln firing temperatures sustained at 1350 °C for 24–48 h batch firing will have altered significantly the original chemical composition of the starting materials, such as china clay, potash, steatite, calcined bone ash, soapstone (steatite), flints and alabaster.

Wollastonite

Formulated as $CaSiO_3$, or $CaO.SiO_2$, this is calcium metasilicate; formed at elevated temperatures between 800 and 1000 °C in the kiln from the reaction between calcareous components in the porcelain paste mixture such as bone ash and the silicaceous china clay and soapstone, wollastonite itself is unstable above 1200 °C, when it forms *pseudowollastonite*, a chemically polymorphic, hexagonal crystalline form. Hence, at the operating temperatures of most kilns adopted for the production of soft paste porcelains, namely 1300–1400 °C, calcium metasilicate will be in the pseudowollastonite form. Although made synthetically in the kiln process, wollastonite does occur naturally in several locations and is often associated with lime, garnets and diopside. In the presence of magnesium and iron, and especially with diopside, a calcium magnesium metasilicate $CaMg(SiO_3)_2$, and enstatite $MgSiO_3$, both of the pyroxene group of silicates, wollastonite can form a solid solution at these high temperatures which analyses as $5CaO.2MgO.6SiO_2$, in which molecules of akermanite have been identified, $Ca_2MgSi_20_7$. When dissolved in a silica glassy matrix, pseudowollastonite can crystallise in a monoclinic form, when it is known as *rivaite*, reminiscent in nomenclature of the ancient Egyptian glassy pigment *cuprorivaite*, $CaCuSiO_3$.

Whitlockite

Formulated as a calcium orthophosphate, $Ca_3(PO_4)_2$, this is derived at high temperatures from the dehydroxylation of hydroxyapatite, $Ca_3(PO_4)_2.Ca(OH)_2$, the primary product of calcination of bone at lower temperatures to make "bone ash". Although present in many places as a natural mineral, especially when combined with fluorine and chlorine as fluorapatite and chloroapatite, written as $3Ca_3P_2O_8.CaF_2$ and $3Ca_3P_2O_8.CaCl_2$, it appears that porcelain manufacturers preferred the use of calcined bones to avoid contamination with the chloride and fluoride radicals in the silicate glassy matrix Also, there is a mineralogical equivalent termed "bone phosphate", of economic importance and formulated as $Ca_3P_2O_8$, which might possibly have been considered as a potential replacement for bone ash but the mineral version seems to possess an rather indefinite composition (Dana, *Textbook on Mineralogy*, 1955) which would not have endeared its potential usage in porcelain paste composition by manufacturers as the quality of the calcium phosphate component in bone ash was critical for the good translucency of the finished articles.

Mullite

Found only at high temperatures and especially in porcelain manufacture, formulated as the aluminium silicate Al_2SiO_5, and derived from silicates such as kyanite, andalusite or sillimanite present in clays, all of which can be written as a combination of alumina and silica, such as $Al._2O_3.SiO_2$.

Sanidine

This is the high temperature stable form of the orthoclase feldspar minerals found above 900 °C, which are formulated as $KAlSi_3O_8$ and examples of which include microcline. Sanidine is found to have incorporated calcium and sodium through high temperature reactions with calcium and sodium ions forming the derivatives species anorthite, $CaAl_2Si_2O_8$, and bytownite, $(Ca,Na)(Al,Si)Si_2O_8$.

Quartz

This perhaps is the most versatile and diverse descriptive label which covers several important components of a porcelain paste mixture: *alpha*-quartz is pure silica, SiO_2, whereas silicates are often represented as a combination of metal oxides and silica, such as sodium metasilicate, Na_2SiO_3, which is often given as $Na_2O.SiO_2$ and forsterite a, magnesium silicate belonging to the olivine family of formula, Mg_2SiO_4, often expressed as $2MgO.SiO_2$. As with most other materials involved in the process of manufacture of porcelain in kilns operating at high temperatures, however, even the simple "silica" itself undergoes structural changes to tridymite and finally to *beta*-cristobalite, the high temperature stable variety, all having the chemical formulation of SiO_2. The situation is compounded by the involvement of different silicaceous materials which comprise silica in varying degrees, such as flint glass, flints, chert, simple silicates, feldspars containing one or more metallic elements and Dillwyn even mentions the addition of smalt to his porcelain body mixture recipe, which is a coloured glass, containing cobalt oxide aluminium oxide and silica and often simply and incorrectly formulated as $CoSiO_3$. The reason for his inclusion of smalt is not clear but could well be attributed to its blue colouration being effective in removal of traces of yellow being otherwise evident in the fired porcelain body As we have seen, structurally quartz has many possibilities for silicon-oxygen chemical coordination through Si–O, O–Si–O and Si=O bonding to create thereby a vast three-dimensional glassy matrix which confers strength and translucency upon the finished porcelain after firing. It is interesting to note here that William Billingsley apparently did not use a glass cullet or frit in his paste mixture, although he does specify the incorporation of high quality flints, which are essentially pure SiO_2, to make a "frit", which has previously been interpreted to mean incorrectly that he did use a glass frit! On the other hand, Lewis Dillwyn in the manufacture of his finest duck-egg porcelain specifically did use flint glass as well as flints in his paste frit preparation. The secret seems to reside in the fineness of preparation and especially of the grinding of this "frit" and not whether heavy flint glass had been incorporated. The selection of almost colourless flints seemed to be highly important as the geological family to which flints belong, including chert, actually has a range of material which is coloured by metal oxide impurities to give essentially different minerals such as jasper, agates, chalcedony, carnelian, amethyst, onyx, opal and touchstone. All of these minerals, needless to say, have the general chemical formulation of SiO_2. Over the temperature range of the firing kilns, hexagonal trapezoidal—tetrahedral alpha-quartz which is the low temperature, stable form of silica converts to the hexagonal trapezoidal-hemihedral form, beta-quartz above 573 °C, and above 870 °C this converts to tridymite, which finally forms *beta*-cristobalite around 1250–1300 °C (Dana, *Textbook of Mineralogy*, 1955). Clearly, a key protocol in future analyses of porcelains will need to differentiate between the silicas of various types added to the mixture and the identification of the glassy silica component in flint or crown glass.

The major chemical components of the porcelain paste recipes have been considered here, but other minor components mentioned in notebooks and work books such as arsenic or antimony oxides, borax, lime, alabaster, alum and gypsum have not been expounded further. The important message from this section is that it is facile to consider in great depth the analytical data obtained from elemental analyses or of stylised metal and non-metal oxides, such as, silica, magnesia, soda and alumina, and in even more complexity, phosphoric acid and its derivatives because of the multiplicity of sourcing of these entities in relation to their possible components in the starting materials. A supporting method is therefore needed to provide analytical data from the molecular species associated with the actual components in the fired porcelain and that is precisely what is going to be explored here.

4.3 The Coal *Versus* Charcoal Dilemma

The alternative usage of coal and charcoal in kilns fired for porcelain manufacture has not really been examined rigorously in the literature thus far, yet it is suspected that this parameter could have been critical in the development of soft paste porcelains that were fired at high temperatures approaching 1300–1400 °C. Lewis Dillwyn has commented in his diaries that the presence of certain materials were detrimental to the appearance of the final porcelain product, especially calcium carbonate, which was responsible for the generation of bubbles through the emission of carbon dioxide gas at temperatures between 650 and 700 °C. Both William Billingsley and Lewis Dillwyn were aware of the problems which could arise from the presence of carbonaceous matter in raw materials during their preparation of their paste mixtures: clearly, at the frit formation stages. the production of gases from the thermal decomposition of organic matter at relatively low temperatures would not have been problematic as any voids in the inorganic matrices would have probably been rectified by the fine grinding processes that were adopted subsequently to create the powder used in the final pastes comprising the porcelain body mixtures. In the production of the bone ash from calcined ox or cow bones, the final heating to temperatures in excess of 1000 °C would certainly have been sufficient to decompose any organic residues remaining after the initial low temperature thermal treatment. It is interesting to compare here the preparation technology of "bone black" or "ivory black", which were highly esteemed as pigments in mediaeval scriptoria for providing a beautiful rather glossy and deep black pigment favoured by scribes for their historiated manuscripts and carbon black inks for application to vellum and parchment manuscripts. Contrast this requirement for a lustrous black product with that of Billingsley and Dillwyn, who demanded a pure white product for their best quality "bone ash" for Swansea and Nantgarw porcelain production: both syntheses depended upon the heating of bones or ivory, the former to produce a deep black end-product, whilst the latter requirement was for a pure white product. The key lies in the temperatures adopted for the heating of the pulverised bones either for black pigment manufacture or for white bone ash production. Whereas for the black pigment manufacture, the bones were heated in a limited supply of air at low temperatures, probably around 300–400 °C, so the carbon produced by the thermal decomposition of the keratotic organic components such as collagen, in the bone ash production, which was accomplished at much higher temperatures in a free flow of air, the carbon produced in the lower temperature thermal decomposition of organic materials burned off to leave the essentially pure white inorganic matrix of calcium triphosphate. An interesting observation made by contemporary historians of bone black production for carbon-black based inks is that the best quality black inks were made using carbons that had the resinous or tarry material present, which assisted the adhesion of the ink to the vellum substrate, which was usually addressed with the addition of gum arabic to the aqueous suspension of carbon particles in water in provision of the fluid inks for scriptoria (Edwards, *Ancient Inks: A Forensic Historical Perspective*, 2015). In fact, scribes were known to prepare their own special black inks not from bones or ivory, but rather from peach or apricot stones which on heating gave rather more organic resinous by-products that were beneficial to the ink adhesion compared with the standard bone black varieties. The excellence of *Indian ink* preparations for use on mediaeval manuscripts has been attributed to the preparation of carbon black from pine trees and the resultant content of resinous material which assisted in the adherence of the ink to the vellum substrates.

4.4 The Importance of Grinding in the Manufacturing Operations

Although perfection in visual appearance was the prime objective in porcelain manufacture, for several reasons outlined above this could not be universally achieved in practice. Faults resulting from the kiln firing processes at high temperatures were often encountered and these contributed to the excessively large kiln losses experienced in Nantgarw and to a lesser extent at Swansea: incorrect control of temperature gave rise to "sagging", wherein the distortion of the shape of the porcelain piece was aggravated to tender it being unusable. Other firing faults occurred resulting in smaller distortions in flatware shape and in the appearance of hairline cracks called firing faults. Often the latter especially were deemed acceptable and were glazed over. Other surface blemishes and faults occurred as pits and bubbles which again could be filled in with glaze and later masked using shrewdly placed motifs such as insects or flower buds during the enamelling or gilding processes. A more troublesome and unsightly blemish occurred when foreign material was manifest as particulate matter or stains at or near the surface: its origins are speculative and range between foreign matter introduced in the differential sourcing of the paste component materials, organic contamination of the bone ash, cinder formation from charcoal-fired or coal fired kilns and ineffective grinding and possible contamination of the starting materials. It has already been mentioned that William Billingsley was particularly keen on the preparation of his starting components, sourcing only the finest materials, and especially on the grinding of his bone ash frit for which he used the services of a local miller, no doubt under strict supervision. The concept of cross contamination is well known and appreciated in analytical science: indeed the eponymous *Locard's Exchange Principle* proposed by Professor Edmond Locard over a hundred years ago is the mainstay of forensic analysis today and can be stated as "Every contact leaves a trace". What this means is that whereas the actual grinding operation of the bone ash or of the later porcelain mixture also known as "frit" undertaken at Nantgarw or Swansea could be monitored closely, what is imponderable is the extent to which the milling grindstones were cleaned before this was undertaken. The millstones or querns were normally made from millstone grit, a very hard quartzite material, or even granite and came in various sizes from simple handheld devices to larger units powered by horses or cattle, and later on by water or steam. It is quite possible, therefore, that during the fine grinding process, which could have taken many hours to complete, particulate matter was transferred from the mill stones themselves to the porcelain frit or bone ash, especially if the mill had been in use previously for grinding hard materials which had eroded their contact surfaces. Unlike modern analytical mills which adopt agate or tempered steels for the grinding mechanism, the millstone grit stones would have had rather rough and pitted surfaces which would have collected detritus from previous operations and even after superficial cleaning this could easily have then contaminated the porcelain frits and their components in subsequent use. A relevant factor in this discussion is the relative hardness of the milling grindstones used and the components of the porcelain frit and paste: essentially, the harder the components of the porcelain frit, especially after its initial firing process and preparatory to its secondary grinding operation, then the longer it will take to effect a finely ground composite and the more likely it is that contamination from particulate matter will occur.

A scientific measure of the relative hardness or softness of materials is provided by Mohs Scale of mineral hardness, devised by Friedrich Mohs in 1812: this is a qualitative ten-point scale which characterises the scratch resistance of minerals based upon the ability of a harder material to scratch a softer one. It is much used by geologists today to identify types of natural rocks and mineral deposits. It is an ordinal scale, which means that it is more akin to a geometric progression, meaning that corundum is only one unit higher than topaz on the Mohs Scale but it is twice as hard in absolute terms. A listing of the standard minerals on the Mohs Scale is included in Appendix C and several others have been included because of their particular relevance to porcelain manufacture matched alongside a numerical listing of their absolute hardness. We can immediately see that the range of hardness possessed by the various components of porcelain pastes is large and this alone amplifies the importance of the consideration of the grinding process in our discussions here. For example, talc, alabaster and china clays are very soft materials whereas quartz sand and feldspar are much harder, perhaps by a 100 X or so. The ability, therefore, of a millstone to effectively grind a combination of these materials to the same degree of fineness is a tall order indeed and should not be under estimated. Depending upon the precise nature of the millstones used in the grinding device, the hardness of millstone grit rock and granites is estimated at about 6–7 on the Mohs Scale.

(See Appendix C)

In this way, the incorporation of unlikely contaminants into the porcelain mixtures would have occurred and perhaps given rise to particulate blemishes on the porcelain surfaces. This factor does not seem to have been addressed hitherto but it should be recognised analytically: the next section describes a novel analytical investigation of such particulate contamination of an otherwise perfect piece of Nantgarw porcelain, a finished and locally decorated dessert plate.

4.5 Faults Observed on Porcelain Surfaces

The observation of a few dark blemishes on a Nantgarw dessert plate during analysis of the body was investigated to ascertain if a chemical identity for these could be established which might indicate their source or origin in the manufacturing process: this procedure was facilitated by the use of a laser beam to interrogate the specimen in situ. It is believed that this has not been attempted hitherto because this would necessarily have entailed considerable damage to the specimen by physical, mechanical removal of the object concerned. Here it was possible to analyse these objects under the glaze through a translational adjustment of the laser beam focus.

4.5.1 Dark Blemishes on Porcelain Surfaces

In this context, an adverse consequence to the presence of carbonaceous material in the kiln during the firing process now becomes apparent, aside from the generation of carbon dioxide gas as the carbon burns off in the kiln as highlighted earlier. The collector of fine porcelains from the 18th and 19th Centuries is soon made aware of the blemishes that occur on finished porcelain pieces, which have still been beautifully gilded and decorated for sale—in some cases, illustrated here, the blemishes which are in the form of pits, bubbles in the glaze or of black particles, have been masked by random decorative motifs such as insects, moths, butterflies and single flower buds inserted by the ceramic artists during their decoration work. An example of this practice can be seen in this text in the large Nantgarw platter shown in Fig. 4.2 and of the disfiguring surface blemishes in an otherwise perfect Nantgarw dessert plate shown in Figs. 4.3 and 4.4. This is not unique to Swansea and Nantgarw porcelains and examples are to be found on Derby and Pinxton porcelains from the Billingsley period (*ca.* 1790–1800) as well as on porcelains from other factories: another example of such a blemish is shown in Fig. 4.5, a Pinxton porcelain cup and saucer enamelled by William Billingsley in John Coke's factory there in 1798. What has not seemingly been addressed in the literature thus far is the nature of these surface blemishes: generally black or very dark brown in colour, they are particulate and they occur at or very close to the surface of the porcelain on or sometimes in the glaze itself. It is reasonable to propose, therefore, that the particulate matter was transferred onto the otherwise perfect biscuit porcelain base at the glazing stage: should this disfigurement have occurred prior to this and have been identified actually within the biscuit or unglazed piece itself then the porcelain would have surely been rejected as being unsuitable unless the blemish could be easily masked perhaps by the addition of a small insect or flower during the decoration process. We need to examine, therefore, what could have occurred in the glazing process, in the "glost" kiln, to create particulate matter which would be transferred onto the porcelain, which at that stage may have been glazed in the white or possibly have been enamelled, gilded and treated with the glazing slip recipe for lower temperature

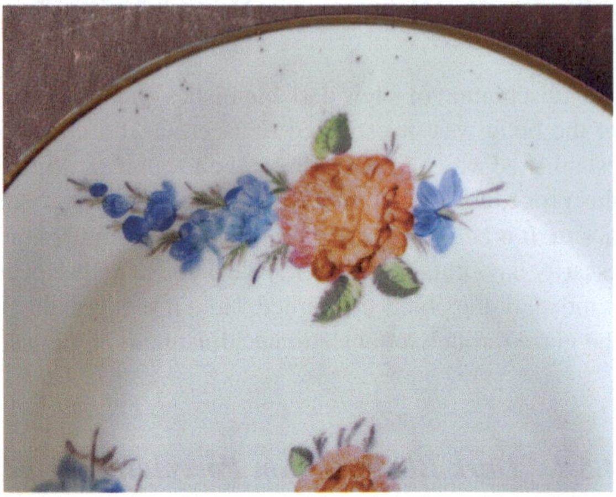

Fig. 4.3 Nantgarw dessert plate, locally decorated with sprays orange roses and blue delphiniums and simple edge gilding. Marked NANT-GARW C.W. impressed on base. Note the profusion of black spots, pits and blemishes which it is theorised would have rendered this otherwise perfectly translucent and well-shaped plate unacceptable for the London market

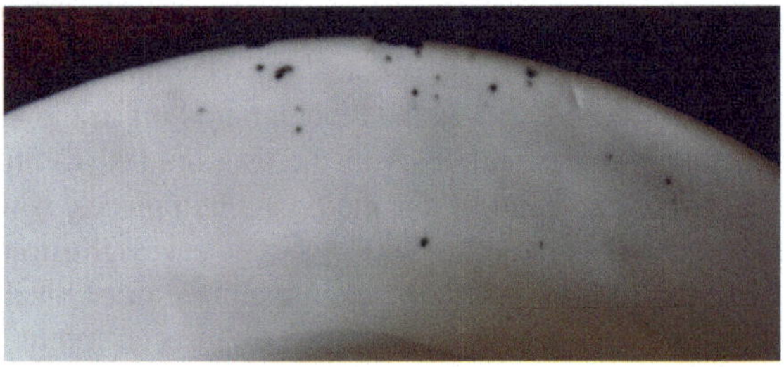

Fig. 4.4 Reverse side of the Nantgarw dessert plate shown in Fig. 6.28, where a range of blemishes, black spots, assorted lumps and pits occur near the edge

firing. This hypothesis is supported by the example illustrated earlier of a Pinxton cup and saucer, shown in Fig. 4.5, where two blemishes occur on the surface of the saucer close to the rim and one inside the bottom of the cup, which under close examination under high magnification can be seen to comprise a black particle encased in a bubble under the glaze. This is very interesting scientifically because it suggests that volatile material has been produced from the particle in the glost kiln and has failed to escape the surface tension of the viscous glaze, creating a bubble in the glaze by partial thermal decomposition, leaving a residual black particle. A possible explanation of this phenomenon would be that organic material lodged between the glaze and the biscuit porcelain substrate has been decomposed thermally in the kiln process and has released a vapour, perhaps carbon dioxide or methane, leaving behind a residue of carbon. The gas under low pressure has not been able to escape through the viscous prototype glaze layer and, therefore, has created a definite bubble, at the bottom of which we can still see the residual black particle.

We should now attempt to explain the origin of these black particles and their contamination of their host decorated biscuit porcelain in the glazing process. Although the presence of soot particles in the main firing kiln would be anticipated to have been burned off at the operating kiln temperature of 1300 °C or higher, becoming carbon dioxide, particularly in the oxidising atmosphere of the Nantgarw

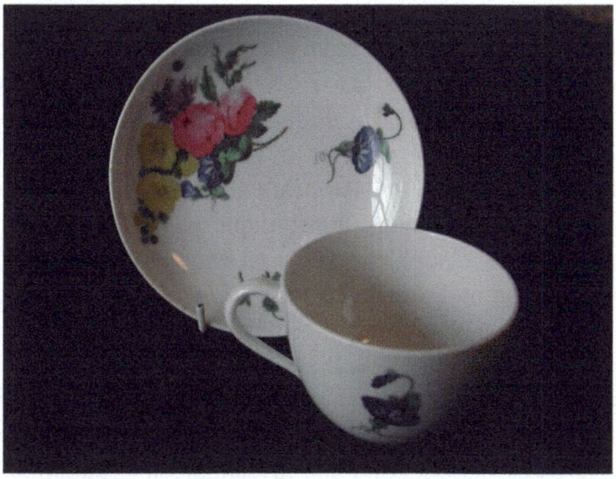

Fig. 4.5 Pinxton porcelain tea cup and saucer, decorated by William Billingsley, ca. 1798, extremely rare and ungilded

kiln (John, Nantgarw Porcelain, 1948) the situation in the lower temperature glost kiln would have been significantly different especially if soot particles were being deposited on the wet glazed porcelain—and at the lower temperatures of the kiln these would possibly not have been volatilised. The apparent indiscriminate use of charcoal, wood and coal for firing the kilns could only have created a further complexity in the formation of particulate soot in this context: carbon normally burns at temperatures between 700 and 1000 °C in a plentiful supply of oxygen, whereas charcoal is manufactured by burning wood at a temperature of between 300 and 350 °C in a limited supply of air, enough to partially combust the organic molecules in wood, such as lignins and resins, whilst preserving significant amounts of elemental carbon to which they were thermally degraded. Both coal and charcoal, therefore, contained significant residues of organic resins but coal-fired furnaces burned hotter and thereby destroyed the resins, producing soot in the process; in contrast, charcoal-fired furnaces could still produce sooty deposits from resin combustion but these would have been better preserved at the lower kiln temperatures. A related comment derives from the use of charcoal in Indian inks used in antiquity, which in admixture with water and gum arabic gave resinous inks much admired by the mediaeval scribes in scriptoria because of the superior adhesive powers of the ink on the vellum. The hypothesis, here, therefore, is that the use of charcoals and indifferent quality coals, along with wood in some cases, could have materially assisted in the production of particulate matter at the glazing stages of porcelains and thereby contributed to blemishes of the type shown on the underside of the locally decorated Nantgarw plate shown in Fig. 4.4, which is otherwise perfectly formed. Although obviously imperfect enough to warrant its withdrawal from sale to the London distributors, as an otherwise perfect survivor from kiln wastage and damage through sagging and warping, which affected 90% of the Nantgarw output from the kilns, nevertheless, it passed muster and was sent for sale locally.

This discussion is not merely conjectural and there is historical documentation and correspondence in several contemporary porcelain manufactories, although not for Swansea or Nantgarw, which indicates the extent of the apparently indiscriminate usage of wood, charcoal and coal for the firing of ceramics kilns (Young, *English Porcelain, 1745–1795: Its Makers, Design, Marketing and Consumption*, 1999; Anderson, *Derby Porcelain*, 2000). The Staffordshire potters favoured coal firing for their kilns, using locally sourced coals of sometimes varying quality and burning efficiency: this appears rather haphazard and whilst one manufacturer used coal exclusively, other family members in the factory business used wood—factories such as Caughley used coal exclusively for kiln firing because of their close association and their proximity to local ironworks industries in the Severn Valley. Despite this, wood seemed to be the fuel of choice for early porcelain manufacturers, probably in accord with the established wood-fired Chinese kilns and the finest continental porcelain factories such as Meissen and Sevres were wood-fired too. In fact, Meissen only adopted coal-fired kilns as late as 1839 and Sevres in 1845. Anderson (*Derby Porcelain*, 2000) has investigated this situation for Derby in particular but has also cited some correspondence from William Billingsley whilst he was operating the china factory at Pinxton, noting the shift to coal fuel for firing porcelain kilns towards the end of the 18th Century, paralleling its adoption as the fuel for steam engines in the emerging Industrial Revolution. Reference has already been made to fuel consumption as a significant "hidden cost" in porcelain production: William Billingsley has recorded the figures for a weekly production of Pinxton porcelain in 1795 as 4.5 tons of coal for the biscuit kilns and 3 cords of wood (10 tons) for the glazing kilns, which extrapolates to an annual consumption of 312 tons of coal and 500 tons of wood, adding some 12% to the cost of the raw materials for preparation of the set pieces of porcelain—excluding the ancillary heating required on site for the pretreatment and processing of raw materials for the porcelain production. William Billingsley's purchase figures quoted exclude the carriage cost, which actually more than doubled the final cost for delivery of the material to the factory site. So, it is easy to appreciate the concern of porcelain manufacturers at the "hidden cost" of production of their wares and Billingsley's obvious selection of Nantgarw as a site for his start-up venture there, sitting atop the South Wales coalfield and its inexhaustible supply of highest grade steam coal, renowned for its thermal output and calorific value with low ash content. There has been a suggestion that Billingsley and Walker chose the site at Nantgarw for their early "first phase" venture into porcelain manufacture because it was "isolated", and that they wished to hide away in secret upon leaving Barr, Flight & Barr at Worcester to facilitate this enterprise. This surely cannot be tenable when it is realised that Billingsley and Walker actually departed from Worcester amicably and even left with a parting gift of cash from Martin Barr which they used to set up kilns in Nantgarw. The attitude of Martin Barr towards Lewis Dillwyn when the news of their engagement at Swansea to manufacture porcelain there was effectively a warning that Billingsley and Walker should remember their agreement that they would not divulge porcelain recipes or secrets to a third party as agreed upon their departure from Worcester, but this did not specifically and legally prevent them from undertaking porcelain manufacture themselves!

In the application of Raman spectroscopic techniques to porcelain analyses described later, it will be relevant to attempt to study the composition of these blemished or black particles situated near the surface of finished porcelain pieces in an attempt to understand their origin.

References

J.A. Anderson, *Derby Porcelain and the Early English Fine Ceramic Industry*, Ph.D. Thesis, University of Leicester, UK, Oct 2000

J.W. Anthony, R.A. Bideaux, K.W. Bladh, M.C. Nichols (eds.), *Handbook of Mineralogy, Volume VII: Silica and Silicates* (Mineralogical Society of America, Chantilly, Virginia, USA, 1995)

A.H. Church, Cantor Lectures on Some Points of Contact Between the Scientific and Artistic Aspects of Pottery and Porcelain, Lecture IV. *Journal of the Society of the Arts* (1880, January 14), pp. 126–129. (Extended in monograph by Trounce Publishers, London, 1881)

Sir A.H. Church, in *English Porcelain: A Handbook to the China Made in England During the 18th Century as Illustrated by Specimens Chiefly in the National Collection.* A South Kensington Museum Handbook (Chapman & Hall Ltd., London, 1885 and 1894)

E.S. Dana, *Textbook of Mineralogy: An Extended Treatise on Crystallography and Physical Mineralogy*, 4th edn. (revised by W.E. Ford, J. Wiley & Sons, Inc., New York and Chapman & Hall Ltd., London, 1955)

L.W. Dillwyn, *Notes on the Experimental Production of Swansea Porcelain Bodies and Glazes.* Made by Lewis Weston Dillwyn with Samuel Walker at the Swansea China Works Between 1815 and 1817. Presented to the Library of the Victoria &Albert Museum, South Kensington, London by John Campbell in 1920. (Reproduced in Eccles & Rackham, *Analysed Specimens of English Porcelain*, 1922, see reference)

H. Eccles, B. Rackham, *Analysed Specimens of English Porcelain in the V&A Museum Collection* (Victoria & Albert Museum, South Kensington, London, 1922)

H.G.M. Edwards, Ancient Inks: A Forensic Historical Perspective, in *Encyclopaedia of Scientific Dating Methods*, eds. by W.J. Rink, J. Thompson, A.J.T. Jull, J.B. Paces, L. Heamann (Springer, Heidelberg Germany, 2015), pp. 48–52

M. Hillis, The Development of Welsh Porcelain Bodies, in *Welsh Ceramics in Context, Part II*, ed. by J. Gray (Royal Institution of South Wales, Swansea, 2005), pp. 170–192

W.D. John, *Nantgarw Porcelain* (Ceramic Book Co., Newport, 1948)

E. Morton Nance, *The Pottery and Porcelain of Swansea and Nantgarw* (B.T. Batsford Ltd., London, 1942)

E.G. Ramsay, W.R.H. Ramsay, in *Bow: Britain's Pioneering Porcelain Manufactory of the 18th Century*, The International Ceramics Fair & Seminar, Park Lane Hotel, London, 16 June, 2007a, pp. 1–16

W.R.H. Ramsay, E.G. Ramsay, A Classification of Bow Porcelain from First Patent to Closure: c.1743–1774. Proc. R. Soc. Vic. **119**(1), 1–68 (2007b). ISSN 0035-9211-1-168

J. Taylor, *The Complete Practical Potter* (Shelton, Stoke-upon-Trent, 1847)

M.S. Tite, M. Bimson, A technological study of english porcelains. Archaeometry **33**, 3–27 (1991)

W. Turner, *The Ceramics of Swansea and Nantgarw* (Bemrose & Sons, Old Bailey, London, 1897)

H. Young, *English Porcelain, 1745–1795: Its Makers, Design, Marketing and Consumption* (Victoria & Albert Museum Publications, South Kensington, London, 1999)

Chapter 5
Molecular Composition of Porcelain Bodies from Modern Microanalytical Studies

Abstract Detailed discussion of the molecular composition of porcelain bodies as determined from modern microanalytical studies using scanning electron microscopy and energy dispersive X-ray scattering: this differs from the earlier gravimetric or wet chemical methods used in that the composition of microdomains is interrogated rather than the bulk material. The advantage is that only small amounts of material are required for the analysis. However, the determination of lime, alumina, silica, magnesia, and phosphate still requires the correlation with formulation of starting materials as novel minerals formed at high temperatures are identified: enstatite, sanidine, cristobalite, leucite andbytownite. Mention made of first vibrational spectroscopic studies of porcelains—a molecular analysis for correlation with the elemental analyses.

Keywords SEM · EDAXS · Instrumental microanalysis · High temperature mineral formation · Molecular · Elemental

Compositional data from early wet chemical analyses are generally reported in terms of oxide materials such as silica, alumina, magnesia, phosphate and lime—chemically, these are represented by the formulations SiO_2, Al_2O_3, MgO, PO_4^{3-} and CaO—which are all derived from the dissolution, separation and precipitation reactions used in their extraction, and this aspect has already been discussed. Useful though these quantitative data are for the differentiation between the pastes, mixtures and additives used by different manufactories, the real molecular composition of the original, intact porcelain bodies is rather different; the question then arises—how does one access this analytical information about the materials actually present in the fired porcelains, which should give a better discrimination facility for the different types of porcelain bodies, their kiln firing temperatures and post-firing treatment. Later, modern analytical data are of the elemental compositional variety, which is especially useful for a consideration of material microdomains in the solid state and from which an idea of the chemical composition of the fired bodies can

H. G. M. Edwards, *Nantgarw and Swansea Porcelains*,
https://doi.org/10.1007/978-3-319-77631-6_5

be gained. By this means, materials such as enstatite, cristobalite, calcite, feldspar, tricalcium phosphate, anorthite, whitlockite, mullite, plagioclase, leucite, gypsum, corundum, steatite, bytownite, sanidine and *alpha*-quartz have all been identified in Nantgarw and Swansea porcelains (Owen et al. 1998). Clearly, the discovery of such a wide range of compositions and materials from scanning electron microscopy analyses enables much information to be gleaned about the kiln procedures adopted by Billingsley and Walker and then by Dillwyn in their quest to achieve high quality porcelains at Swansea and Nantgarw, rivalling the production of porcelain at Sevres in the first quarter of the 19th century.

Vibrational spectroscopy, and in particular Raman spectroscopy, can complement the analytical interpretations afforded by elemental scanning electron microscopy and diffraction techniques by accessing the molecular composition of porcelain bodies and is additionally able to do so non-destructively by confocal imaging through the transparent glaze where access to unglazed regions is not possible. Hence, for the first time, the analyst can study complete, decorated porcelain objects and determine the chemical composition at the micron level without the removal of samples; because the Raman spectroscopic technique gives information on molecular bonds, then a more detailed description of the types of silicate can be forthcoming and even a definition of the coloured pigments used in the decoration, which may themselves be factory relevant and specific. An example of such an exercise has been the assignment of a fine inlaid porcelain mahogany tea table (Fig. 5.1) to the Royal Rockingham factory around 1835 by Raman spectroscopic correlation of the pigments and porcelain body found on the table with that for a red-griffin marked Rockingham dinner plate (Edwards et al. 2004). It is of interest to compare the Raman spectroscopic results for the porcelain bodies and pigments used in the Rockingham factory with those of a similar study undertaken by Colomban et al. on porcelains from the Sevres factory in the 18th and 19th centuries (Colomban et al. 2001, Colomban and Treppoz 2001).

Table 5.1 gives a summary of the minerals that have been identified using elemental SEM/EDAXS analysis of soft paste porcelains from several 18th and 19th century factories such as Bow, Derby and Worcester along with Nantgarw and Swansea, including their chemical formulae and descriptions, compiled from the available analytical literature (Tite and Bimson 1991; Morgulis and Janacek 1931; Owen et al. 1998; Edwards et al. 2004). Minerals found to occur in Nantgarw and Swansea wares have been identified further with an * in this Table. Several important points can be made from this Table: firstly, specific materials and mineral phases (such as high temperature polymorphs of *alpha*-quartz and silicates from the pyroxene and feldspar series) are evident which have been formed under different kiln firing conditions which are much more discriminative than the simple Ca:P or similar elemental ratios and silica compositions that have been derived from earlier wet chemical extraction processes. Secondly, the presence of mineral phases which are different but which yield similar or identical elemental ratios, such as *alpha*-quartz and cristobalite or apatite and whitlockite, are not recognised in the earlier wet chemical extractions but are clearly differentiated in the later elemental microchemical and X-ray diffraction studies. This is particularly important for the potential discrimination between porcelain factories,

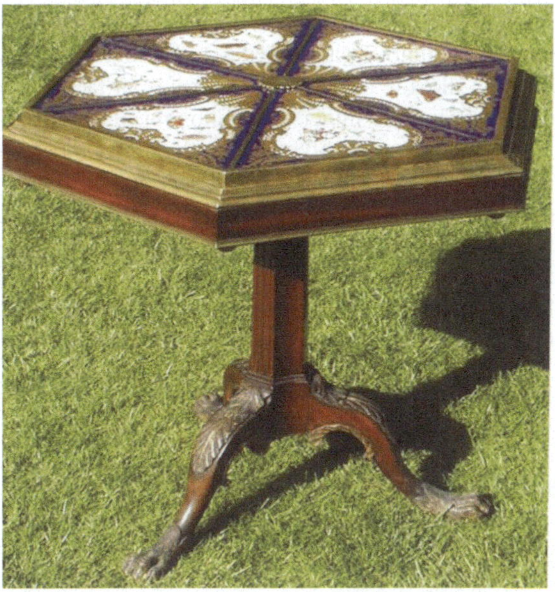

Fig. 5.1 An 18th century mahogany tripod table inlaid with Rockingham porcelain panels bound in brass and highly decorated with fine floral painting on a blue ground, ca. 1830–1840, from the sale of effects at Wentworth Castle, South Yorkshire, seat of the Fitzwilliam family. Along with their relative, Earl Fitzwilliam of Wentworth House, they were enthusiastic patrons of the Rockingham china works and several unique items in porcelain were commissioned from the Rockingham factory, including this table, porcelain garden furniture and the famous large "Elephant" vases. The table top porcelain was analysed using non-destructive Raman spectroscopy, which provided a match for the porcelain body with a Royal Rockingham period marked plate from ca. 1835 and with similar enamelled pigments, therefore placing it firmly as a unique Rockingham porcelain item. Reproduced courtesy of Bryan Bowden, Esq

especially when experimental paste mixture compositions were being trialled along with kiln temperatures; in this respect, the identification of relatively small amounts of lead- containing components ascribed to the use of flint glass frit additive in the Nantgarw and Swansea porcelains is especially crucial as Eccles and Rackham (*Analysed Specimens of English Porcelain*, 1922) and Church (*English Porcelain*, 1894) had concluded that the presence of lead in their analyses was attributed to the interference and contamination of the glaze on their specimens. Of course, this could not be the explanation for the SEM results of Owen et al. (1998, 1999), who used exclusively factory glazed wasters, except for the one specimen analysed of finished, decorated porcelain from the Biddulph service. Hence, the presence of lead in the mineral phases noted by Owen et al. is entirely reconcilable with the use of varying amounts of lead glass frit adopted by Billingsley and Walker in their experiments in Swansea, although it appears that this was not a component at Nantgarw, and the Eccles and Rackham interpretation of its presence as interference from the glaze contamination must therefore be treated with circumspection.

The identification of titanium in small amounts by SEM analysis itself generates an intriguing explanation: although attributed to contamination in the raw products of processing, titanium dioxide in the anatase form is in fact a minor component in kaolin, and this has been used hitherto as a monitor of kiln firing temperatures, since the conversion of the anatase polymorph to the high temperature stable rutile form occurs around 850 °C. This has been suggested hitherto as a useful spectroscopic marker for the so-called large dragon kiln temperatures in Chinese porcelain manufacture of the Ming dynasty at Hangzhou, although recent work on the thermal interconversion of anatase and rutile by will undoubtedly necessitate some reappraisal of several conclusions made in the literature regarding kiln firing temperatures on the sole basis of the presence or otherwise of anatase.

5.1 Summary of Perspective on the Use of Analytical Data for the Identification of Porcelains

It is appropriate to present here a summary of the main features which characterise the adoption of chemical analytical data for the identification of porcelains and, in particular, the basis offered for the discrimination of unknown porcelain specimens and their assignment to specific factories.

- The major difference between the older chemical analyses and modern analyses occurs as a result of the range of elemental and molecular instrumentation that can now be brought to bear to assess the macroscopic and microscopic composition of ceramic specimens. Whereas the earlier wet chemical analyses of a century ago, such as those exemplified by Eccles and Rackham in 1922 are still useful precursors of the equivalent modern nano-scale investigations, the information realised from both types of analyses are essentially incompatible unless several important corrections are applied as has been highlighted in the discussions above. The early chemical analysts expressed their determinations in terms of chemical species which actually bore no real correlation with the species actually present in the porcelain after its high temperature firing process treatment in the kiln. Simplistic representations of metallic content as metal oxides were at best subjective—such as calcium oxide, silica, magnesia, soda and potash, which in reality were derived from a host of related and more complex chemical compounds as illustrated in Table 5.1, such as mullite, whitlockite, hydroxyapatite, anorthite and steatite. Modern instrumental analyses, especially data from SEM/EDAXS and XRF measurements allied with XRD, IR and Raman spectroscopic studies have indicated the true complexity of ceramic materials.
- Perhaps the most critical difficulty emerging for an objective comparison between analytical data emanating from the multitude of porcelain factories which existed between the mid-18th and the mid-19th Centuries is the reliability of the determinations matched against the changes invoked to experimental porcelain bodies

Table 5.1 Compilation of materials found in 18th and 19th century porcelains (Edwards 2017)

Name	Brief description	Chemical formula
*Alpha -quartz	Silica	SiO_2
*Apatite	Tricalcium phosphate	$Ca_3(PO_4)_2$
Hydroxyapatite	Calcium hydroxy phosphate	$Ca_5(OH)(PO4)_3$
*Cristobalite	High T quartz polymorph	SiO_2
*Tridymite	High T quartz polymorph	SiO_2
Enstatite	Magnesium silicate	$MgSiO_3$
*Whitlockite	Anhydrous hydroxyapatite	$Ca_3(PO_4)_2$
*Calcite	Calcium carbonate	$CaCO_3$
*Mullite	Aluminium silicate	Al_2SiO_5
Feldspar	Potassium aluminium silicate	$KAlSi_3O_8$
Sanidine	High T alkaline feldspar	$KNaAlSi_3O_8$
Corundum	Alumina	Al_2O_3
Pyroxene	Calcium magnesium inosilicate	$CaMg(Si,Al)_2O_6$
Talc	Hydrated magnesium silicate	$Mg_3Si_4O_{10}(OH)_2$
Muscovite	Mica phyllosilicate	$KAl_2(Si_3Al)O_{10}(OH,F)_2$
Microcline	Low T alkaline feldspar	$KAlSi_3O_8$
*Magnesia	Magnesium oxide	MgO
Forsterite	Magnesium olivine	Mg_2SiO_4
Kaolinite	Aluminium silicate	$Al_2Si_2O_5(OH)_4$
*Lime	Calcium oxide	CaO
Rutile	Titanium oxide	TiO_2
Anatase	Titanium oxide	TiO_2
Haematite	Iron oxide	Fe_2O_3
Soda	Sodium oxide	Na_2O
Soda ash	Sodium carbonate	Na_2CO_3
*Potash	Potassium carbonate	K_2CO_3
Wollastonite	Calcium silicate	$CaSiO_3$
Gypsum	Calcium sulfate dihydrate	$CaSO_4.2H_2O$
Anhydrite	Calcium sulfate	$CaSO_4$
*Anorthite	Calcium aluminium silicate	$Ca_2Al_2Si_2O_8$
*Bytownite	Plagioclase feldspar	$(Ca,Na)[Al(Al,Si)Si_2O_8]$
Albite	Sodium aluminium silicate	$NaAlSi_3O_8$
*Leucite	Potassium aluminium silicate	$KAlSi_2O_6$
*Steatite	Magnesium silicate	$Mg_3Si_4O_{11}$
Flint glass cullet	Lead potassium silicate (var. comp.	$(PbO)_{0.8}(K_2O)_{0.45}(SiO_2)_{4.4}$

*Indicates species present in Nantgarw and/or Swansea porcelain bodies which may or may not have been adequately documented during the manufacturing phase. Factories and their porcelain products were in a highly competitive situation and improvements were being made to the basic bodies and glazes to retain market superiority and a cutting edge over rival concerns. Hence, when Twitchett (*Derby Porcelain 1748–1842*, 2002) writes authoritatively about Derby porcelain analyses during the earliest Chelsea-Derby period from 1770, through the William Duesbury I and II periods marked with a puce crown, batons and cursive D up to 1800, and finally the Robert Bloor period with cursive red mark or stencilled crown and concentric circles mark up to 1835/40, then we are tempted to think of only three identifiable bodies being adopted over perhaps some 70 years production. Few analyses have been undertaken to date to test this hypothesis, which means that the analytical figures expressed by Twitchett for Derby porcelain are assumed to be definitive. This means that without access to specific notes from the Duesburys or Bloor regarding paste composition changes during their ownership of the Derby China Works, we are tempted to assume that a constant composition recipe was used for many years operation and this surely is highly unlikely. The question can then be posed as to the reliability of discrimination that can be effected between rival factories operating in similar periods: put simply, how confident can we be in discriminating between Coalport, Worcester, Derby, Swansea and Nantgarw pieces in the 1815–1820 chronological period if we are uncertain of the changes made to paste composition, which may be subtle or otherwise?

- A third problem lies in the accuracy and precision of description of the analytical data that is extant in the existing literature for analysed porcelains: the accuracy of an analytical result can be expressed in terms of a numerical value with an estimated error bar based upon several determinations made from the same or identical samples—such as, a calcium oxide percentage of 10.2 ± 0.1%, which represents a 1% error evaluated from actual measurements—the more of these that are undertaken on a specimen gives a better reliability of determination, or accuracy, through a lower error bar. The precision of a measurement on the other hand expresses how close the experimental determination is to the real value: for example, if Lewis Dillwyn noted that for a particular porcelain body he advocated mixing 10.1% calcium oxide into his paste body then the analytically accurate figure stated above is also a very precise one as it has estimated the compositional presence of the calcium oxide in this sample to within one standard deviation of the experimental error. Analysts can express their determinative values as confidence intervals, so for example, an accurate value of 10.1 ± 0.1% calcium oxide with a three-sigma confidence interval means that the precise value will lie to within the range 9.8–10.4%, with a 95% confidence interval. This means of course that it is almost impossible to state **exactly** what the percentage of analytically determined calcium oxide is within any specified sample, but we can safely predict realistically and with some certainty that it will lie somewhere between 9.8 and 10.4%. This means, of course, that the analytical determinations of this same component estimated from samples of different factories may at first seem to be sensibly different but, in reality, they may overlap within the three-sigma criterion: hence a Coalport specimen determination of calcium oxide which results in an analytical value of 9.7 +/– 0.1% is actually indistinguishable from that of the Swansea determination as its three-sigma range lies between 9.4 and 10.0%. This is not generally appreciated by non-analytically oriented observers but it can really mean that the elemental or molecular entity that is specifically chosen for demonstrating that two factories had a different component composition must be selected very carefully indeed.

- A second point that needs to be considered carefully in analytical work of this nature is the variance that is to be encountered in sampling for what are otherwise essentially identical specimens or pieces. This means that we should not assume that every item of porcelain manufactured in a batch has precisely similar component percentages—and what is even more serious is the question as the whether or not there is a significant change expected between batches—the so-called batch variation—which can be caused by systematic and random errors invoked in making up the components of a paste. Inaccuracies in weighing at the factory, in the use of batches with greater levels of impurities and consequently smaller proportions of active constituent, in the use of calcined materials which have been subjected to varying degrees of drying or hydration, all contribute to incipient analytical errors that affect the reliability of the final determinations. It is, of course, impossible to quantify or even assess the importance of such imprecise departures deduced from the standards specified in a recipe work book, but the final precision of any analytical data will naturally incorporate these deviations. For example, Lewis Dillwyn in his work book recipes where he undertook experimental changes to his porcelain bodies dating from 1815 to 1817 cites a compositional recipe for his new "*trident*" porcelain as "*Body No. 2: silica 4 parts, soaprock 1 part, potash ½ part*". The accuracy of making up such a composition from its component parts is not specified but for a fifty-five lbs aliquot of powdered mixture in the weights and measures operating at that time, he would require 40 lbs silica, 10 lb of soaprock and 5 lbs potash: hence, only a ½ lb error in weighing each component would thus produce incipient errors of 1.3, 5, and 10%, respectively, in the silica, soaprock and potash contents which would be translated over into the final analytical determinations of calcium oxide alumina, magnesia, potash and soda. This would also be reflected in apparent changes to the analytical determinations of these elemental oxides internally between specimens, as an inter-specimen variation, which would not necessarily be indicative of subtle and intended predetermined changes made to the initial paste compositions themselves. This could explain several "inconsistencies" noted in the analytical determinations reported for many factories as will be seen in Table 5.2, which gives the collected analytical data from some reported experiments from specimens originating from different china manufactories in the mid- to late-18th and early 19th Centuries.

Table 5.2 Collected analytical data for chemical analyses on english and welsh porcelains

Component %										
Factory/Date	SiO_2	Al_2O_3	CaO	H_3PO_4	MgO	Na_2O	K_2CO_3	Fe_2O_3	PbO	References
Bow 1750/60	40.00	16.00	24.00	17.30	0.80	1.30	0.60	Trace		Church (1894)
Bow 1750/60	43.58	8.36	24.47	18.95	0.60	1.20	0.85		1.75	Eccles and Rackham (1922)
Bow 1750/60	42.80	8.84	28.32	18.10	trace	1.12	0.72		0.50	Eccles and Rackham (1922)
Bow 1750/60	55.10	16.5	15.12	11.50	0.40	0.31	0.70	Trace		Eccles and Rackham (1922)
Bow	50.38	7.78	24.87	13.66	trace	0.70	0.53		1.49	Eccles and Rackham (1922)
Chelsea 1760	40.20	8.40	27.40	20.30	trace	1.00	0.90	1.20		Church (1894)
Chelsea 1765	45.52	12.06	26.00	14.27		0.73	0.93			Eccles and Rackham (1922)
Lowestoft	41.60	19.14	10.8	25.81	1.22	1.02	0.41			Spelman (1902)
Bristol 1760	69.96	24.43	1.50	0.17	trace	1.92	1.36		1.50	Eccles and Rackham (1922)
K'angHsi 1700	71.82	23.04	0.63	0.19	trace	2.12	1.89		0.60	Eccles and Rackham (1922)
New hall 1790	70.14	24.26	1.92	0.41	trace	2.91	0.62			Eccles and Rackham (1922)
New hall 1790	73.56	19.30	4.02	0.24	trace	2.10	0.92		0.67	Eccles and Rackham (1922)
Worcester 1790	70.43	4.82	4.64		10.5			4.65		Eccles and Rackham (1922)

(continued)

Table 5.2 (continued)

Component %										
Factory/Date	SiO_2	Al_2O_3	CaO	H_3PO_4	MgO	Na_2O	K_2CO_3	Fe_2O_3	PbO	References
Worcester 1800	75.36	18.87	2.81	0.16	0.18	2.00	1.27			Eccles and Rackham (1922)
Chelsea	64 76	6.00	25.00	0.23	trace	1.82	2.58		0.55	Eccles and Rackham (1922)
Chelsea	69 10	5.90	20.50		0.71	0.69	3.30	Trace		Eccles and Rackham (1922)
L'ton H 1760	76.16	4.30	9.28						6.50	Eccles and Rackham (1922)
L'pool 1760	62.76	4.45	20.10	2.10	0.18	1.14	1.16		8.58	Eccles and Rackham (1922)
Derby 1785	41.94	15.97	24.28	14.96	0.20	1.06	0.90		0.36	Eccles and Rackham (1922)
Derby 1790	43.20	12.40	25.32	12.40						Eccles and Rackham (1922)
Pinxton 1796	41.94	13.80	24.80	14.10	0.62	0.78	0.59		4.02	Eccles and Rackham (1922)
Coalport 1820	42.88	15.06	23.16	16.30						Eccles and Rackham (1922)
Swansea 1817	81.56	8.90	0.70	0.33	4.26	0.61	3.27	Trace		Eccles and Rackham (1922)
Swansea 1817	84.00	8.67	0.97		2.50	0.60	2.91			Eccles and Rackham (1922)
BFB Worc. 1810	76.10	7.55	1.30		9.65	1.40	4.06			Eccles and Rackham (1922)
	76.44	7.68	1.78	0.41	6.14				5.40	Eccles and Rackham (1922)

5.1.1 The Interpretation of the Analytical Data Cited in Table 5.2

Several important conclusions can be made regarding the analytical data for porcelains manufactured from the mid- to late-18th and early 19th Centuries accumulated in Table 5.2 which reflects directly on the ability of analytical instrumentation to discriminate between different factories for the identification and attribution of unknown or suspected pieces. Firstly, in the cases where the analytical data for more than one item from the same factory in a specific chronological period have been reported, it is seen that there is a "spread" of component % ages which requires some interpretation and explanation. Firstly, close examination of the chemical compositional data in Table 5.2 reveals that the % ages of the metal oxides do not indicate the starting compositions per se of the components in the starting paste mixture recipes: as has been stated above, only phosphorus can be relied upon to directly reflect the amount of bone ash used in the recipe, and even then, because of the chemical formulation errors in the composition of calcined bone ash which have been perpetuated in later literature interpretations, the amount of bone ash deduced from the phosphorus determinations as pentoxide, phosphate ions or phosphoric acid, itself is often rather arbitrary. In all other cases, the determination of the elemental percentages cannot directly be translated into specific components that may be present in the recipes: calcium, for example, occurs in calcined bone ash, lime, alabaster and clays. Likewise, sodium and potassium occur in feldspars, potash and soda, glass cullet and in clays. Silicon occurs in quartz sand, clays, feldspars and soapstone. Magnesium is a major constituent of feldspars and lead of glass cullet, only when this is flint glass, otherwise sodium and potassium are expected to occur also from this source. Where dolomitized limestone has been substituted for calcite or aragonite, the presence of elemental magnesium is thereby increased.

What can be revealed from Table 5.2 are several important points which can be used to interpret discrete compositions from the analytical data: this is best achieved from noting the presences and absences of some of the key experimentally determined metal oxides:

- The presence of phosphorus exclusively indicates the use of calcined bone ash in the recipe: the quantitation of the composition is best then achieved using the table given here which corrects for potential formulaic miscalculations in the early literature. A test of the correctness of this procedure is seen in the back-calculation and estimation of calcined bone ash determined from the % ages of phosphorus oxides, phosphates and phosphoric acid obtained by wet chemical and instrumental analyses—when it can be observed that the figures match closely the formulation given for Swansea duck-egg china in Dillwyn's favoured recipe cited in his notes. Conversely, the absence of phosphorus can be conclusively reasoned to exclude the presence of any bone ash in the recipe—as found for example in Chinese and also in European hard paste porcelains.
- Lead in porcelain recipes equally derives from only one source, namely, the presence of flint glass cullet ground in the preparatory frit: crown or soda glass contains

no lead, so the absence of lead in itself does not indicate the absence of glass frit as several factories favoured the lighter so-called soda-glass (lead-free) cullet in their recipes.

- Magnesium arises from the minerals feldspar and steatite, soapstone (petuntse), so a high magnesium content will invariably be indicative of the incorporation of soapstone in the recipe, which was usually added to strengthen the porcelain matrix by replacing kaolin and to make the paste more fusible at higher kiln temperatures—effectively, therefore, reducing the kiln temperatures by as much as 200 °C and providing more kiln temperature control.
- Iron usually occurs in the porcelain analyses from its presence in sand as haematite—the finest, white quartz sand is relatively free from iron contaminant but other sands have a variable content. Iron compounds can confer a rather yellow colour on fired porcelains and thus Dillwyn found the need to add smalt, a blue cobalt aluminosilicate pigment, to offset this yellow colouration and produce a clear, white transparency.
- Sulfur traces arise from the use of gypsum and alum content in the porcelain recipe additives. Other additives include borax, sodium tetraborate $Na_4B_4O_{10}$, to aid the fusion of alkaline flux in the recipes, but generally the boron content is not specified; arsenic oxide was also added, with an unknown purpose.
- Silicon oxide cannot be specifically related to particular components in any recipe as silicon occurs in several major components, including soapstone, china clay, glass frit, and river sand. Nevertheless, it is apparent that several porcelain formulations had very high silica content, such as Swansea trident porcelain.
- Aluminium also is non-specific for the analytical attribution of recipe components since it arises from feldspars and alum.

Following the discussion of the Raman spectral data obtained from specimens of several porcelains studied in this exercise, we shall revisit these conclusions and assess the matching of the molecular spectral data with the presence or otherwise of the individual recipe components, where these have been stated in factory records. Another important statement must be made here regarding the differences which appear to occur in the % age compositions of the metal oxides which seemingly reflect real compositional changes in the porcelain mixtures. One needs to be careful in this approach as it should be apparent that factory owners regularly experimented with changes in recipe and formulation of their paste and where this information has been recorded for posterity, as in the case quoted here for Lewis Dillwyn's trial recipes cited in Appendix 1, this does not necessarily mean that a new porcelain mixture was finally put into full production. It is quite possible that only a few trial pieces were ever made and possibly after firing these were destroyed and deposited in the factory waste dump—hence, evidence of a "new commercial porcelain body" obtained from analyses of single wasters from factory dumps is not in itself forensic proof that a new body had been put into full production commercially. Such has been the allegation and conclusion from some analytical studies in the past and a word of caution needs to be exercised in this context. Hence, the announcement of a new Swansea body, for example from the analysis of a single waster or shard is not sufficient to promulgate this statement as a fact, and one should be naturally aware of the consequences: thus, it is even more relevant and important to gather precise analytical information from complete, finished porcelain pieces, which have made the transition from kiln to point of commercial sale to verify their actual production status.

In this context, it is worth mentioning a little-known fact that could enter into an analytical interpretation: Anderson (*Derby Porcelain*, 2000), in her comprehensive and detailed thesis on Derby porcelain manufacture has accessed a wealth of correspondence and records which had only been cursorily examined hitherto. In the 1790s, Anderson found a record that William Duesbury had purchased from a London supplier a large quantity of French and Chinese hard paste porcelain shards, amounting to "several tons"—his purpose in doing this is unclear and it has been suggested that he wanted to examine the shapes and styles of the porcelains being supplied by his competitors, but surely it is mooted that he could better achieve this through purchase of exemplars from the factory retail suppliers of perfect pieces? As it happens, when his successor at Derby, Robert Bloor, visited the London atelier of Robins & Randall and saw James Plant painting Nantgarw plates, he purchased six or so in number, from which he then modelled very closely his own Derby service for Lord Ongley (Edwards, *Swansea and Nantgarw Porcelain: A Scientific Reappraisal*, 2017; Morton Nance, *The Pottery and Porcelain of Swansea and Nantgarw*, 1942). A better explanation for the purchase of this large tonnage of shards, which is supported by a later commentary, is that Duesbury was experimenting with the addition of ground hard paste porcelain cullet to his Derby frit, by which process he hoped to achieve a better strength and resilience for his porcelain body, as already exhibited in the Chinese porcelains, for example. This would be a reasonable avenue to test by ongoing experiments; it is not reported if his experiments were successful, but in view of the large quantity of hard paste shards purchased there may be a significant number of Derby pieces circulating which have a very unusual body composition, as yet undetected by analytical interrogation! The second point to be noted here is that if the proposed experiments were unsuccessful then a large quantity of shards of non-Derby origin would have been deposited in the waste dump(s) on the factory premises—giving credibility to the assertion made elsewhere in this text that it would be facile to assume that all shards found at a manufactory site actually originated from the factory processes being undertaken there and may in fact have been "bought in" from external sources. In selecting a shard for analysis, therefore, it is essential that it has either the factory mark applied (a rather rare occurrence) or that it matches a piece of marked and finished porcelain sold from the factory. Clearly, this is not always possible, and the analysts should therefore be aware of the consequences of forming conclusions relating to novel porcelain body compositions from single chards of indeterminate origins which may even have found their way into waste factory material from later attempts to continue manufacturing at the site . A case in point in this context is at Nantgarw, where William Pardoe did successful manufacture earthenware clay pipes for several decades from about 1835, some twelve years after the closure of the Nantgarw China Works. Waste material from Pardoes's operations could, therefore, be reasonably mistaken for Nantgarw (Billingsley & Walker) china shards unless excluded on the basis of shape, which may not be immediately apparent for very small fragments.

References

J.A. Anderson, *Derby Porcelain and the Early English Fine Ceramic Industry*, Ph.D. Thesis, University of Leicester, UK, October, 2000

Sir A.H. Church, *English Porcelain: A Handbook to the China Made in England During the 18th Century as Illustrated by Specimens Chiefly in the National Collection*, (A South Kensington Museum Handbook, Chapman & Hall Ltd., London, 1885 and 1894)

P. Colomban, F. Treppoz, Identification and differentiation of ancient and modern European porcelains by Raman macro- and micro-spectroscopy. J. Raman Spectrosc. **32**, 93–102 (2001)

P. Colomban, G. Segon, X. Faurel, Differentiation of antique ceramics from the Raman spectra of their coloured glazes and paintings. J. Raman Spectrosc. **32**, 351–360 (2001)

L.W. Dillwyn, *Notes on the Experimental Production of Swansea Porcelain Bodies and Glazes.* Made by Lewis Weston Dillwyn with Samuel Walker at the Swansea China Works Between 1815 and 1817. Presented to the Library of the Victoria &Albert Museum, South Kensington, London by John Campbell in 1920. (Reproduced in Eccles & Rackham, *Analysed Specimens of English Porcelain*, 1922, see reference)

H. Eccles, B. Rackham, *Analysed Specimens of English Porcelain in the V&A Museum Collection* (Victoria & Albert Museum, South Kensington, London, 1922)

H.G.M. Edwards, *Swansea and Nantgarw Porcelain: A Scientific Reappraisal* (Springer Publishing, Dordrecht, 2017)

H.G.M. Edwards, P. Colomban, B. Bowden, Raman spectroscopic analysis of an english soft-paste Porcelain Plaque-mounted table. J. Raman Spectrosc. **35**, 656–661 (2004)

S. Morgulis, E. Janacek, Studies on the chemical composition of bone ash. J. Biol. Chem. **93**, 455–466 (1931)

E. Morton Nance, *The Pottery and Porcelain of Swansea and Nantgarw*, (B.T. Batsford Ltd., London, 1942)

J.V. Owen, M.L. Morrison, Sagged phosphatic Nantgarw Porcelain (ca. 1813–1820): casualty of overfiring or a fertile paste?. Geoarchaeol. **14**, 313–332 (1999)

J.V. Owen, J.O. Wilstead, R.W. Williams, T.E. Day, A tale of two cities: compositional characteristics of some Nantgarw and Swansea porcelains and their implications for Kiln wastage. J. Archaeol. Sci. **25**, 359–375 (1998)

M.S. Tite, M. Bimson, A technological study of english porcelains. Archaeometry **33**, 3–27 (1991)

J. Twitchett, *Derby Porcelain 1748–1848: An Illustrated Guide* (Antique Collectors Club, Woodbridge, Suffolk, 2002)

Chapter 6
Raman Spectroscopic Studies of Swansea and Nantgarw Porcelains

Abstract This chapter describes the nondestructive molecular vibrational Raman spectroscopic study of specimens of perfect, finished Swansea and Nantgarw porcelain, some shards from the Nantgarw China Works site, and some exemplars of unknown attribution which satisfy many of the requirements of Swansea and Nantgarw porcelain assignment. The three Swansea bodies, comprising trident, glassy and duck-egg paste are easily differentiated within minutes using this technique and a portable Raman spectrometer operating with a probe permits access to the interior surfaces of cups, bowls, jugs, spill vases and ornamental porcelain items. In conclusion, it is stated that a spectroscopic protocol can now be applied for the characterisation of Swansea and Nantgarw porcelains and the identification of unknown specimens.

Keywords Raman spectroscopy · Nondestructive sampling · Finished and perfect specimens unknown exemplars · Protocol identification

It has already been stated earlier that a major problem facing analysts of fine ceramics is the destructive nature of the sampling procedures imposed on the specimen by the adoption of a particular technique: even micro sampling involves the excision of small specimens of porcelain paste for analytical purposes—usually in the form of a small drill boring or of a small chip taken from a masked area such as the base of a footrim. This is perfectly acceptable in the case of shards excavated from a factory site as usually several kilos of material can be supplied in this way. However, in some cases, even this is not considered acceptable—and it is even less so when perfect finished and flawless porcelain items of some rarity need analysis. Hence, it is the situation thus far that the only pieces of Swansea and Nantgarw porcelain which have been analysed are either already badly damaged or broken pieces from a museum collection (Eccles and Rackham 1922), shards (Tite and Bimson 1991; Owen et al. 1998; Owen and Morrison 1999) and a single Swansea plate, believed to be badly cracked, from a named service (Owen et al. 1998).

H. G. M. Edwards, *Nantgarw and Swansea Porcelains*,
https://doi.org/10.1007/978-3-319-77631-6_6

For the first time, the non-destructive analysis of Swansea and Nantgarw porcelains can now be reported using Raman spectroscopy: the very essence of this investigation is a preliminary survey of a range of items from each factory, including several shards from a Nantgarw site excavation which was carried out in the 1990s by the Gwent and Glamorgan Archaeological Trust. These shards were deposited in the Nantgarw china Works Museum, where they are archived today. The novel aspect of this analytical study is that the interrogation of the porcelain body is accomplished by the non-contact laser beam irradiation of pristine examples of Swansea and Nantgarw porcelains from personal collections, accomplished without the necessity for any chemical or mechanical pretreatment, so the specimen is returned to its display case unharmed by the investigative procedure. As can be appreciated, this methodology is a quantum leap forward for analytical science in the porcelain arena and a fundamental objective of this exercise is an assessment of whether or not is it possible to differentiate between Swansea and Nantgarw porcelains, and also some other selected exemplars, objectively using the characteristic Raman spectral signals obtained in these experiments.

6.1 Specimens for Raman Spectroscopic Analysis

The following specimens were provided as complete undamaged and decorated pieces from an established collection, along with shards from the Nantgarw China Works site, Tyla Gwyn, Nantgarw, Wales:

1. Nantgarw shard, moulded plate rim, unglazed biscuit (Fig. 6.1).
2. Nantgarw shard, moulded tazza rim with spherical ball attachments, unglazed biscuit (Fig. 6.1). A decorated unmarked Nantgarw tazza of this shape and style is shown in Fig. 6.2.
3. Nantgarw shard, unmoulded indented dish rim, glazed (Fig. 6.3).
4. Nantgarw shard, plate footrim, glazed (Fig. 6.3).
5. Nantgarw shard, plate rim, glazed (Fig. 6.3).
6. Nantgarw spill vase, London decorated with floral groups (Fig. 1.1).
7. Swansea spill vase, duck-egg porcelain, locally decorated with chinoiserie landscape scene and a dancing Chinaman (Fig. 1.9).
8. Swansea tea bowl, glassy porcelain (Fig. 6.4).
9. Swansea deep soup dish, locally decorated with exotic birds and flowers by William Pollard (Fig. 1.2).
10. Nantgarw coffee cup and saucer, decorated in London by Moses Webster with pink roses (Fig. 6.5).
11. Potential "Swansea platter" attributed to Henry Morris. (Fig. 6.6) with a SWANSEA. red stencil mark (Fig. 6.7).
12. Unknown teacup, London shape, possibly Swansea (Fig. 6.8) with a characteristic duck-egg translucency (Fig. 6.9).

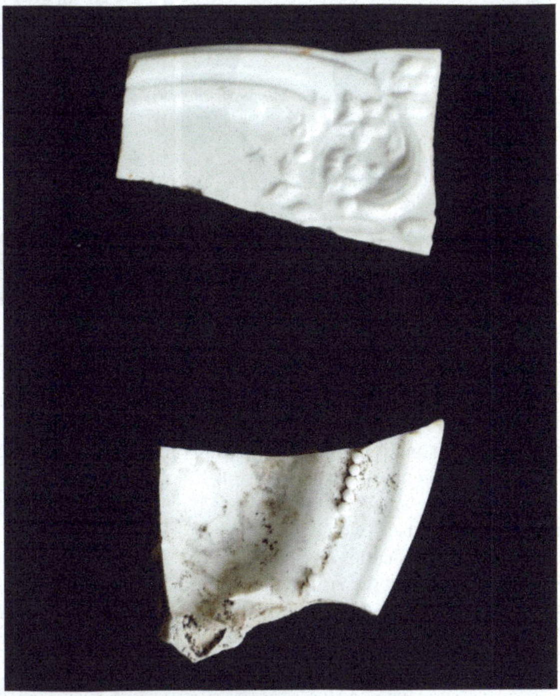

Fig. 6.1 Two Nantgarw porcelain shards, obtained from the excavations carried out at the site of the Nantgarw China Works in the 1990s and since stored in the archive at Nantgarw House, Tyla Gwyn, formerly used as a residence by William Billingsley, 1817–1820. Biscuit unglazed porcelain: the upper shard is from an indented plate rim with characteristic embossed flower moulding with ribbons, and the lower shard is from a small tazza with applied spherical porcelain balls arranged concentrically around the rim. Courtesy of the Nantgarw China Works Trust, Tyla Gwyn, Nantgarw, South Wales

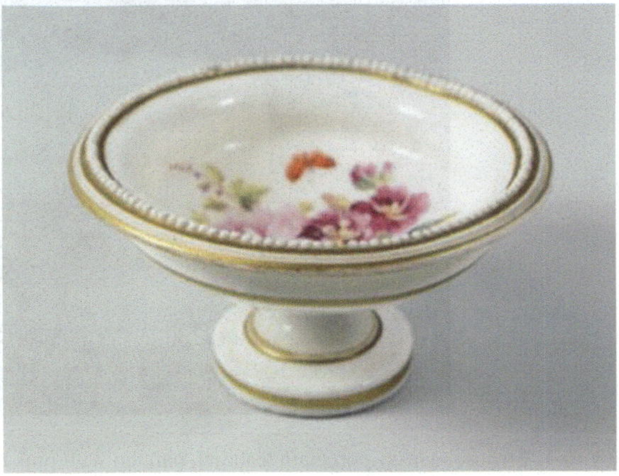

Fig. 6.2 Nantgarw small porcelain tazza, London decorated with flower groups, exhibiting the characteristic applied spherical porcelain balls arrangement seen in the lower shard in Fig. 4.1. Unmarked. Private collection

13. Swansea cup and saucer, duck-egg porcelain, basket weave pattern moulding with cartouche (Fig. 6.10), with characteristic *Swansea* red enamel script mark in William Billingsley's handwriting (Fig. 6.11).
14. Swansea dessert plate, trident porcelain, locally decorated with garden flowers by David Evans, impressed with trident and SWANSEA (Fig. 1.6).

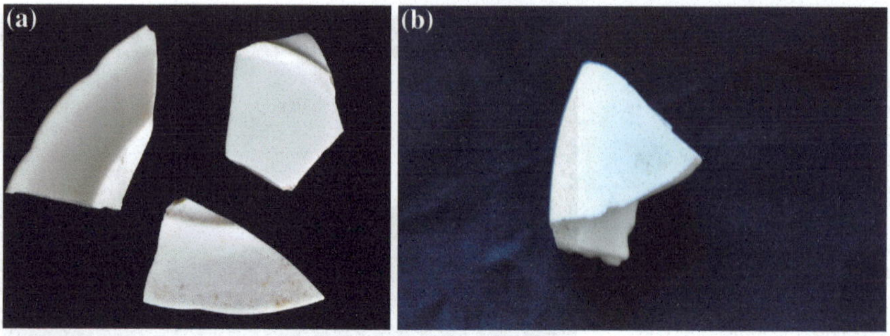

Fig. 6.3 **a** Three Nantgarw porcelain shards, obtained from the excavations carried out at the site of the Nantgarw China Works in the 1990s and since stored in the archive at Nantgarw House, Tyla Gwyn, Nantgarw, formerly used as a residence by William Billingsley, 1817–1820. The upper two shards are from an indented plate rim and footrim, and are glazed whereas the lower shard is from a plain plate rim and is unglazed in biscuit porcelain. Courtesy of the Nantgarw China Works Trust, Tyla Gwyn, Nantgarw, South Wales; **b** Nantgarw porcelain shard selected form the above group excavated at the site of the Nantgarw China Works in the 1990s and since stored in the archive at Nantgarw House, Tyla Gwyn, Nantgarw, formerly used as a residence by William Billingsley, 1817–1820. This shard is an example pf the "sagged" or over-fired porcelain referred to in the text, so-called because of the visual collapse of the moulded shape in the kiln ascribed to an excessively high firing temperature. Courtesy of the Nantgarw China Works Trust, Tyla Gwyn, Nantgarw, South Wales

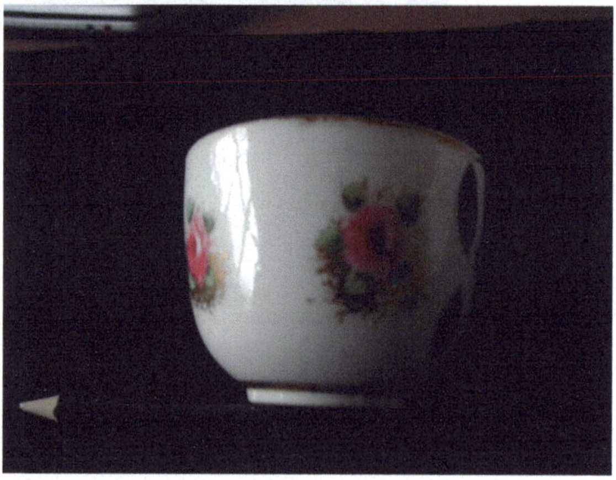

Fig. 6.4 Swansea glassy porcelain teabowl, painted with pink roses on a gilt seaweed ground, ca. 1815–1817, unmarked. Private collection

15. Nantgarw armorial crested dinner plate (Fig. 6.12), glazed but undecorated except for the crest of Phippes in gilt at the rim.
16. Nantgarw locally decorated dessert plate (Fig. 4.4).

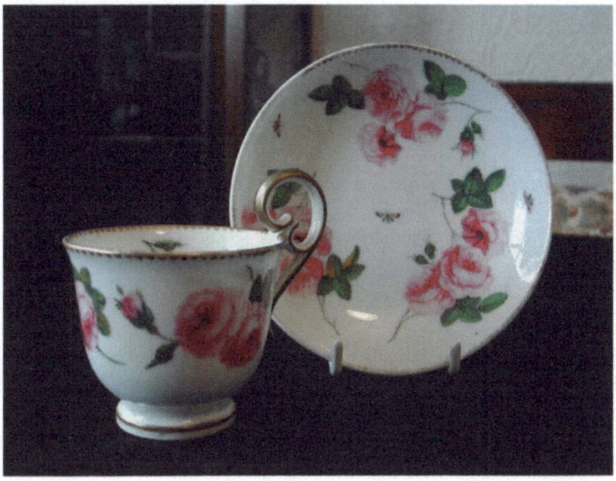

Fig. 6.5 Nantgarw porcelain coffee cup and saucer, heart-shaped handle, London-decorated with groups of pink roses and foliage by Moses Webster in the workshops of Robins and Randall, ca. 1817–1820. Unmarked. The saucer typically has the base unglazed which affords the opportunity to compare analytical spectra form unglazed and glazed areas on the same specimen. Private collection

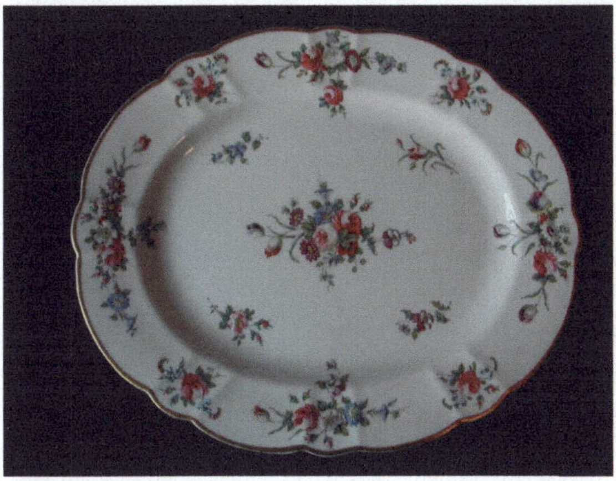

Fig. 6.6 A porcelain meat platter with duck-egg translucency and marked SWANSEA. in red stencil, decorated with floral sprays of tulips, roses and forget-me-nots, attributed to Henry Morris ca. 1820. This is an interesting specimen as it is not a standard Swansea shape reported in the literature but otherwise possesses all the attributes expected of a Swansea porcelain product. Private collection

The shards were provided by the curator of the Nantgarw Museum Trust, Tyla Gwyn, which now is responsible for the maintenance and preservation of the historic site, including the kilns and cottage used by William Billingsley and was rented from Edmund Edwards, the landlord of the site at that time.

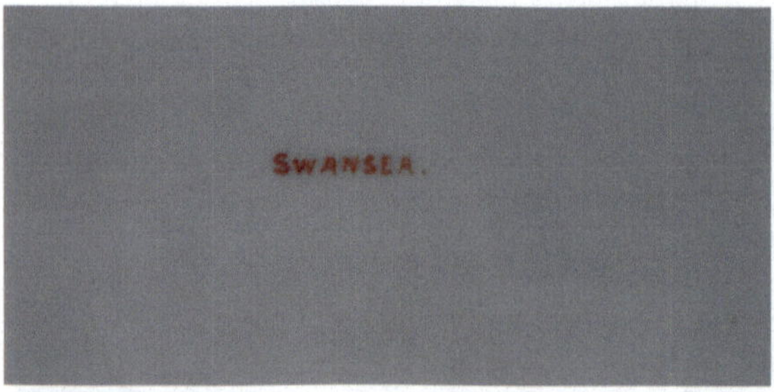

Fig. 6.7 Red stencil SWANSEA. mark on the platter shown in Fig. 6.6, which has been noted previously by Jones and Joseph (Swansea Porcelain, 1988), who ascribed the mark to the local decoration by Henry Morris, especially on porcelain he purchased from extraneous sources that he decorated and fired at Swansea in the years immediately following the closure of the factory, or alternatively to the Swansea porcelain decorated in London for the retailing establishment of Apsley Pellatt in London, where it is sometimes accompanied by their own printed mark

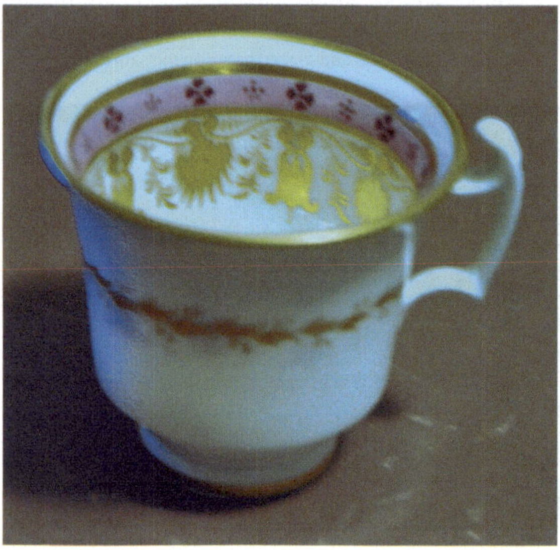

Fig. 6.8 Coffee cup, London shape, which is unmarked and could be Swansea porcelain (see Edwards, *Swansea and Nantgarw Porcelain: A Scientific Reappraisal*, 2017a, b, c, d), and which has several features associated with the factory (see below) including a duck-egg translucency, but the pattern is unrecorded hitherto

6.2 Raman Spectroscopic Instrumentation

The Raman spectroscopic experiments were carried out at the Raman Research Laboratory in the Analytical Laboratories of the University of Bradford personally by the author and Dr Alexander Surtees using a Renishaw RX210 RIAS portable diode laser spectrometer operating at a laser wavelength of 785 nm with a nominal power at

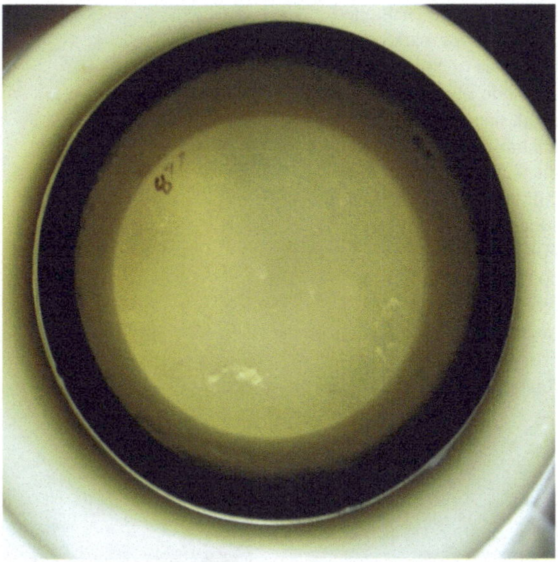

Fig. 6.9 Coffee cup (shown in Fig. 6.27), potentially Swansea, with several attributes of the factory production photographed in transmitted light and exhibiting a duck-egg translucency

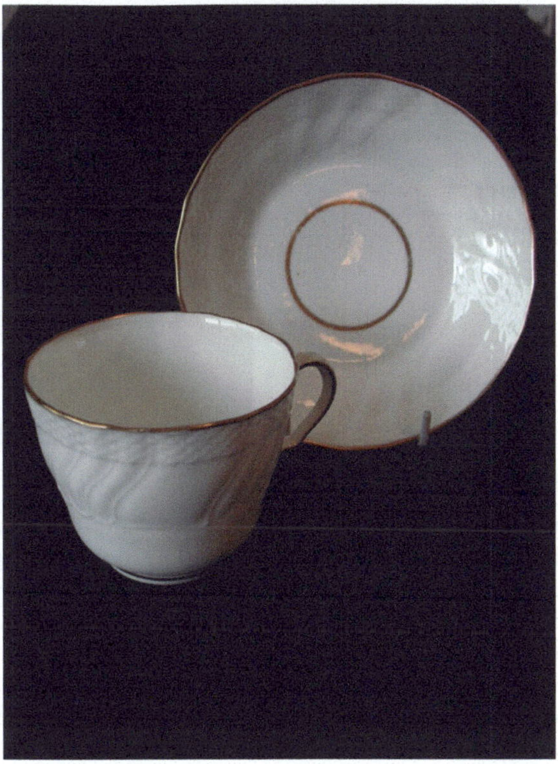

Fig. 6.10 Swansea tea cup and saucer, finest duck-egg porcelain, moulded osier basket weave pattern, simply gilded at the rim and with the double ogee handle, marked Swansea in red enamel cursive script underside both cup and saucer. Private collection

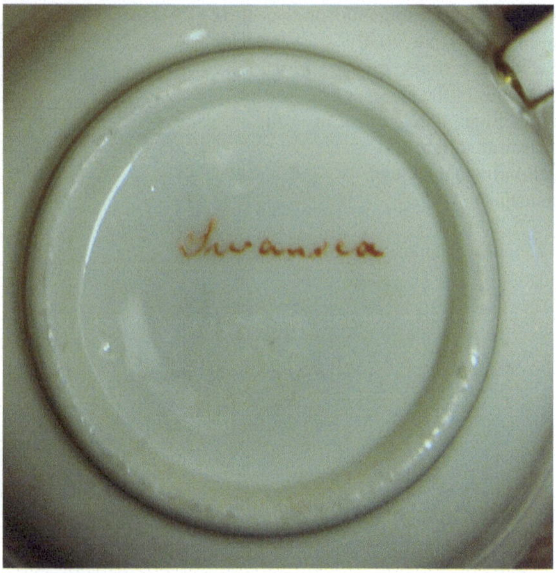

Fig. 6.11 Red enamelled cursive script mark on underside of Swansea saucer shown in Fig. 6.10. This script mark has been unambiguously attributed to the hand of William Billingsley, as identified by Jones and Joseph (*Swansea Porcelain*, 1988) and confirmed by the further researches of Edwards (*Swansea and Nantgarw Porcelain: A Scientific Reappraisal*, 2017a, b, c, d)

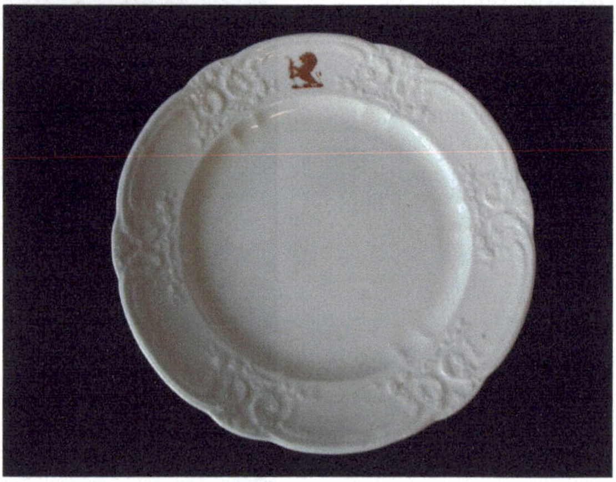

Fig. 6.12 Nantgarw dinner plate, moulded rim, armorial service for the Phippes family, crest in gold with demi-lion passant facing sinister holding a palm frond. Completely undecorated apart from crest. Marked impressed NANT-GARW C.W. ca. 1817–1819

the sample of 50 mW and a spectral resolution of 10 cm^{-1} over a spectral wavenumber range of 100–2000 cm^{-1}. The RX210 was equipped with a 1 m flexible fibre attachment coupled with a 20X Olympus lens objective in a 5:1 arrangement, offering a laser footprint of 100 microns at the sample with a standoff focal distance of 1 cm which facilitates the placement of the probe head for the interrogation of large specimens such as a dinner plate or a platter and the examination of the inside of cups, bowls and spill vases, as will be exemplified here. The flexibility of this arrangement meant that samples of different sizes and shapes could be easily interrogated, from a large flat platter to cylindrical spill vases and small cups, without any contact being made between the probe head and the specimen, which could be mounted vertically or horizontally. After an initial test run on each specimen to determine the correctness of focal imaging, each specimen was examined for 30 co-accumulated spectral scans, each of duration 3 s, and several replicate analyses were carried out for each specimen. Spectral data are presented as obtained either over the whole wavenumber region or as two regions of approximately 1800–1000 and 1050–100 cm^{-1} and appropriate comparisons are effected using spectral stackplots to facilitate the observation of key features and their presence or absence in related specimens.

6.3 Analytical Raman Spectroscopy

Raman spectroscopy is a molecular scattering technique, whereby a monochromatic laser beam is imaged onto the surface of a specimen and the back-scattered Raman radiation is collected and spectrally dispersed on a holographic grating and detected on a charge-coupled detector (CCD) device. The Raman spectrum is displayed as a series of wavenumber shifts from the laser generated Rayleigh line which acts as a wavenumber zero on the wavenumber shift scale Long, *The Raman Effect: A Unified Treatment of the Theory of the Raman Scattering of Molecules*, 2002). Being a molecular technique, the Raman effect gives a unique spectral signature for materials which have chemical bonds, inorganic and organic molecules and molecular ions, and therefore complements elemental detection techniques such as XRD (X-ray Diffraction), XRF (X-ray Fluorescence) and SEM/EDAXS (Scanning Electron Microscopy/Energy Dispersive X-ray Spectroscopy). It is especially valuable for the detection of molecular and molecular-ion components in mixtures and for changes in materials affected during chemical processing, as a result of procedures such as thermal reactions (Edwards 2015a, b). As far as ceramics and porcelains are concerned, the presence of a transparent glaze has no untoward adverse effect since the laser beam can penetrate the glazed layer and interrogate the underlying ceramic body; during this experiment, no damage is done to the sample by the imaged beam of light and no chemical or mechanical pretreatment of the sample is required. The presence of individual minerals or materials is recognised from the observed spectral band wavenumber positions, which can be identified from literature databases. The intensity of Raman scattering is approximately proportional to the individual species concentration, so the greater the concentration of a particular material present the more intense the Raman band observed: not all materials have the same Raman molecular scattering factor cross-section for laser irradiation, however, so several species are always relatively more strongly represented in Raman spectra even when occurring in low concentrations, such as cinnabar, calcite and gypsum (Edwards and Chalmers 2005).

Although the primary investigative instrumental techniques for porcelains and ceramics adopted thus far have been those where the elemental determinations have been paramount, such as XRay Diffraction (XRD), XRay Fluorescence (XRF) and Scanning Electron Microscopy (SEM), the advantage of a molecular technique such as Raman Spectroscopy (RS) is that it readily associates the combination of the chemical elements and their formulation through chemical bond identification: hence, although SEM can correctly provide evidence of mercury, lead, sulfur, oxygen and tin in an artwork or archaeological specimen it is often a matter of conjecture as to how these elements are paired together—for example, potentially these elemental data could signify the presence of the pigments cinnabar or vermilion (mercury sulfide), mosaic gold (tin sulfide), lead oxide (massicot, litharge, plattnerite and minium), lead sulfide (galena), calomel (mercury oxide), cassiterite (tin oxide), and Naples yellow (lead tin sulfide) and several other possibilities too. The heavier metals and anionic entities that are found in the determination, then the more real possibilities are created for their pairing ior association: sometimes, the colour of the particles under investigation will assist in a narrowing down of the possibilities, for example, red might be indicative of cinnabar or litharge, whereas yellow might be indicative of massicot or mosaic gold. In all of these cases, the Raman spectral signatures are different and discriminatory and assist in the evaluation of the correct pairing of metals and their anions: even more problems can arise when one considers the elemental detection of calcium, magnesium, sulfur and oxygen: calcite, aragonite, gypsum, anhydrite, bassanite, magnesite, dolomite, epsomite and hydromagnesite are then all real mineral possibilities but, unlike the pigment examples cited above, all of these are white so the particle colour does not help at all in this case. Again, the RS signatures are discriminatory and can even identify mixtures of these materials, as in dolomitized calcite and in hydrated lime prepared from dolomite.

For ceramics, the presence of silica, silicates and sand, which all contain silicon–oxygen bonds, and Si=O or Si–O bonding combinations in three dimensional networks, give rise to silicaceous matrices which can be quite complex to describe when created at high temperatures. A description of the types of silicate network that can arise at the temperatures adopted for the kiln firing of soft-paste and hard-paste porcelains has been given in specialist texts and a summary has been provided by Edwards (*Swansea and Nantgarw Porcelains: A Scientific Reappraisal*, 2017a, b, c, d), As well as the basic orthosilicates containing the discrete SiO_4^{4-} ions with four-valent silicon from the parent silicic acid, H_4SiO_4, condensed ions such as pyrosilicates and metasilicates also occur and network formation through the sharing of Si=O bonds with neighbouring silicon atoms gives rise to –O–Si–O– linkages in extended three-dimensional structures, conferring strength and rigidity upon the fired porcelain articles. All of these will have different Raman signatures but the resultant spectral band response will be rather broadened by the alternative possible structural conformations of Si=O and Si–O bonding in clays and glasses rather than discrete sharp features as expected for other more crystalline minerals (Ricciardi et al. 2006; Edwards et al. 2004). A major advantage of RS for porcelain analysis is the ability to interrogate the glazed body of a finished article through the glaze, which is transparent to the laser beam, and this opens up the specimen sampling procedure to include perfect examples of rare and museum pieces without the potential of incurring damage, unless the incident laser irradiance (power per unit area, W cm^{-2}) is very large or absorption of the laser radiation wavelength by a pigmented specimen causes localised laser heating. Hence, for the assessment of analytical information obtained from perfect finished porcelain, Raman spectroscopy could provide a solution for the requirement of a non-contact, non-destructive analytical procedure and it is with this in mind that the preliminary analyses of the Nantgarw and Swansea porcelain specimens detailed above have been undertaken—the first on record for Welsh porcelain using this technique.

6.4 Objectives of Raman Spectroscopic Analysis

Several objectives are envisaged in these analytical experiments, namely:

1. Can one obtain definitive RS and key diagnostic Raman spectroscopic signatures for Nantgarw and Swansea porcelain specimens, glazed and decorated?
2. Can one assign the observed Raman features to the paste components of the respective recipes after firing in the kilns and is it possible to correlate the bands observed with the absence or presence of specific components that may have been specified in the paste recipes, viz., the workbook notes of Lewis Dillwyn for Swansea porcelains (Appendix A) and the putative Nantgarw formulation and recipe published by John Shelton (*The Practical Potter*, 1847) allegedly obtained from Samuel Walker before he departed for North America?
3. Does the presence of a glaze interfere with the interpretation of the Raman bands or seriously mask their presence in the interrogation of a porcelain body underneath the glaze?
4. Is it possible to discriminate nondestructively between Nantgarw and Swansea porcelains—the use of genuine, unambiguous, marked specimens can be adopted as standards for the interrogation of specimens of unknown, dubious origin and the resultant exposure of fakes?
5. Was there only a single Nantgarw body composition as has been maintained by many authors, and can one differentiate spectroscopically between the three accepted Swansea porcelain bodies, viz., glassy, duck-egg and trident?
6. With the Raman spectral data available is it then possible to deduce whether or not a piece of porcelain of unknown or merely suspected attribution is actually of Swansea or Nantgarw origin? This is of great value for the scientific attribution of rare or unique items manufactured at Swansea and Nantgarw and their discrimination from forgeries and fakes. This art forensic aspect is an important input to expert opinion based on shape, texture and decoration to provide a holistic assessment of the factory origin.
7. Is there a potential for the detection of "outliers" or associated pieces in large Swansea or Nantgarw services, which have been procured as replacements for broken or damaged items or which have been supplied originally to complete a commission order?

With these objectives and questions to be addressed, the selection of specimens described above taken from an established collection of Swansea and Nantgarw porcelain from the early 19th century have been assembled: these comprise 4 specimens of Nantgarw porcelain, 5 specimens of Swansea porcelain and also 5 representative shards of Nantgarw porcelain from a batch archaeologically excavated from the Nantgarw China Works site in the 1990s and selected from some two kilos of shards provided for this project from the site archive held at the Nantgarw China Works Museum in Tyla Gwyn, Nantgarw. Also, in this section 2 specimens of porcelain which seem to possess the correct criteria for their proper assignment to a Swansea origin in terms of translucency and appearance are included for preliminary assessment to compare with the genuine specimens—a test for the analytical protocols established for the affirmation of a Swansea factory origin. Hence, in this first tranche of specimens, some 16 specimens of Nantgarw, Swansea and putative specimens are taken for non-destructive analysis.

6.5 Basis of Raman Spectral Data Interpretation for Ceramics

The Raman spectroscopic analysis of porcelains primarily addresses the identification of the mineral molecular composition of the bodies and glazes but secondary information can be acquired about the production processes and in particular the temperatures achieved in the kilns in the final stages of porcelain manufacture. Earlier work from Philippe Colomban and his team at the Universite de Pierre et Marie Curie in Paris has established a sound basis for the interpretation of the intricate and highly complex silicate spectra in terms of both the types of silicon–oxygen bonding present as well as the degrees of polymerisation of the silicate matrices which have been subjected to elevated temperatures in kiln fired porcelains (Colomban 2001, 2004, 2005, 2013; Colomban and Treppoz 2001; Colomban et al. 2001, 2004). Glassy silicified ceramics can be classified into five main types, Q^0–Q^4, according to the arrangement of the SiO_4 tetrahedra in their structures, namely:

Q^0: monomeric SiO_4 groups, characterised by SiO stretching vibrations occurring in the wavenumber shift range 800–850 cm^{-1}; an example is forsterite, Mg_2SiO_4;

Q^1: dimeric SiO_4 groups occurring as Si_2O_7 moieties with SiO stretching vibrations in the wavenumber range 1050–1150 cm^{-1}; an example is danburyite, $Ca(BO_3)_2Si_2O_2$;

Q^2: chain silicates comprising Si_3O_9 units with SiO stretching vibrations in the wavenumber range 1050–1100 cm^{-1}; an example is hedenbergite, $CaFeSi_2O_6$;

Q^3: sheet silicates comprising Si_4O_{11} groups with SiO stretching vibrations in the wavenumber range near 1100 cm^{-1}; an example is beryl, $Be_3Al_2Si_6O_{18}$;

Q^4: tectosilicates and SiO_2 structures with SiO stretching vibrations in the wavenumber range 1150–1250 cm^{-1}; an example is oligoclase, $(Ca,Na)(Al,Si)AlSi_2O_8$

Alpha-quartz has a well-defined crystal structure with a characteristic and strong SiO symmetric stretching band assigned to O=Si=O units in the Raman spectrum at 464 cm^{-1}: in fused silica this broadens and is found shifted to higher wavenumbers under mechanical stress—so, for example, in meteorites which have been subjected to high impact stresses the quartz vibrational band shifts to 520 cm^{-1}, characteristic of shocked quartz, termed coesite, In glasses and glazes found on finished porcelains, an estimate of the degree of polymerisation (DP) of the silicate matrix can be evaluated from the relative band intensities of broad features centred at 500 and 1000 cm^{-1}. Generally, this DP is measured as the A_{500}:A_{1000} band intensity ratio which increases with the temperature to which the glaze has been subjected: hence, for soft paste porcelains fired at low kiln temperatures which may typically be around 900 °C the A_{500}:A_{1000} ratio is in the range between 0.8 and 1.1, whereas for hard paste porcelains fired at kiln temperatures of up to 1400 °C, this ratio rises to between 1.5 and 1.7. In the current study the prime intention is not to examine the glazes themselves but to interrogate the porcelain bodies through the glaze coating: also, because the laser wavelength used in our studies is in the near infrared region of the electromagnetic spectrum at 785 nm, this tends to produce luminescence emission bands which appear in the same region as the Raman features. This will be discussed in detail later when the spectral data from the Raman spectroscopic experiments will be collated from a detailed comparative analysis. It is useful at this point to itemise the characteristic Raman spectral bands which are expected to occur for individual minerals and materials which are created in the high temperature kilns during the production of porcelains: these are presented in Table 6.1.

6.6 Raman Spectroscopic Results for Nantgarw and Swansea Specimens

Assuming that it is possible to acquire a Raman spectrum of good quality from the specimens selected for analysis, it is certainly possible to achieve an unambiguous answer to question 3 of those cited above. There is a selection of Nantgarw shards which are unglazed and glazed (specimens 1–5) and also areas of finished porcelain exemplars which are deficient in glaze, such as the Nantgarw saucer of specimen 10 (Fig. 6.5), whose base is unglazed naturally in typical Nantgarw fashion, and areas of the base of the Nantgarw spill vase (Fig. 1.1).

Firstly, the examination of the Raman spectra acquired from two undisputedly Nantgarw items; namely, a London decorated but unmarked Nantgarw spill vase (specimen number 6, Fig. 1.1), and an unglazed moulded tazza dish rim fragment in biscuit porcelain (specimen number 2, Fig. 6.1) from the Nantgarw shards excavated from the waste dump at the factory site, respectively, indicated that the glaze had no influence on the suppression of the Raman spectrum or on the interference in observation of the Raman bands arising from the porcelain body itself. The comparative Raman spectra in the higher and lower wavenumber shift regions for these specimens are shown in Figs. 6.13a, b respectively: it can be seen that the two spectra over the wavenumber ranges 1850–1000 and 1050–100 cm^{-1} are essentially identical. This, therefore, paves the way for the nondestructive acquisition of Raman spectroscopic data from perfect, completed and decorated, glazed porcelain specimens, such as the

Table 6.1 Key Raman spectroscopic band wavenumbers/cm^{-1} for body mineral components in Nantgarw and Swansea porcelains

Component mineral	Band wavenumbers/cm^{-1}
Calcium phosphate	415 m, 960 s
Mullite	300 w, 480 m, 600 br ms, 960 m, 1130 mw
Quartz	206 w, 464 s
Enstatite	350 m, 405 m, 520 w, 675 m, 1015 m, 1036 m, 1088 m
Alpha-Wollastonite	575 m, 988 m
Beta-Wollastonite	635 ms, 972 s
Rutile	144 w, 233 mw, 445 mw, 610 mw
Anatase	143 s, 200 w, 398 m, 517 m, 638 m
Cristobalite	230 mw, 420 m
Carbon	1320 br ms, 1585 br ms
Feldspar	190, 275, 325, 411, 488, 508, 522, 645, 971–974
Haematite	229 m, 299 m, 409 ms, 640 br m, 1320 mw br
Calcite	151 m, 283 ms, 712 mw, 1086 s
Gypsum	412 m, 481 m, 617 mw, 1007 s, 1130 mw
Cassiterite	475 w, 635 s, 776 m, 842 mw
Cerussite	152 mw, 1052 s, 1370 mw, 1477 mw
Microcline	266 mw, 286 mw, 466 br m, 514 s, 813 mw, 843 mw
Muscovite	263 s, 412 s, 633 mw, 702 ms, 753 m
Ilmenite	222 mw, 371 m, 680 m
Forsterite/Olivine	835 ms, 860 s
Fayalite/Olivine	820 s, 852 ms
Serpentine	230, 385, 690
Sanidine	170, 280, 475, 510
Coesite	520
Anhydrite	715, 1018 s, 1133

Nantgarw armorial crested dinner plate, with the crest of Phippes (Edwards, *Swansea and Nantgarw Porcelain: A Scientific Reappraisal*, 2017a, b, c, d), impressed NANTGARW-C. W., shown in Fig. 6.12, the locally decorated dessert plate shown in Fig. 4.4 and the London decorated cup and saucer, unmarked, shown in Fig. 6.5, all beautiful examples of genuine Nantgarw porcelain. A stack plot of the first two, the crested armorial plate and the locally decorated plate are shown in Figs. 6.14a, b for high and low wavenumber regions, respectively, 1850–1000 and 1050–100 cm^{-1}; these Raman spectra are also identical with each other in both wavenumber regions, and the Raman spectrum of the saucer (Fig. 6.5), not shown here, additionally matches both: a molecular assignment of the key features will now be undertaken to assist the recognition of particular components of the porcelain paste composition.

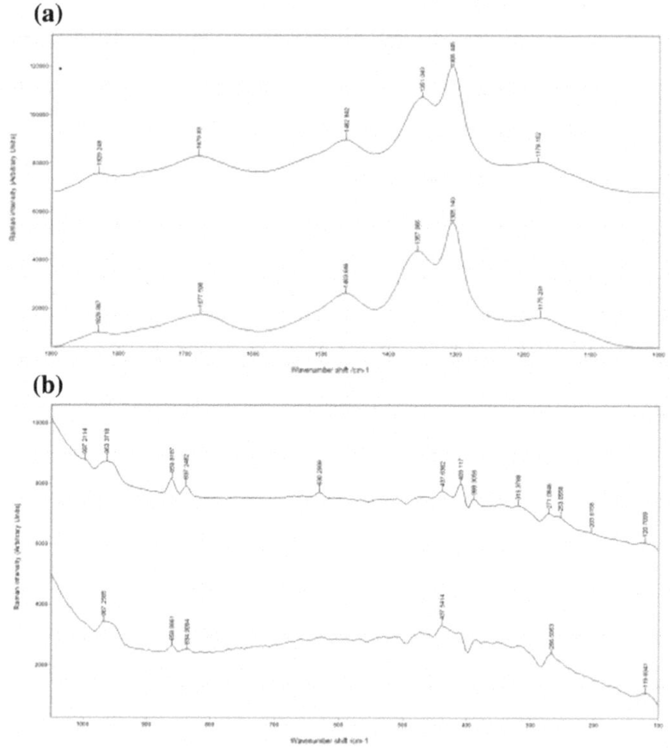

Fig. 6.13 Raman spectral stackplot of Nantgarw cylindrical spill vase (upper spectrum) and moulded tazza shard (lower spectrum): **a** wavenumber region 1850–1000 cm^{-1}; **b** wavenumber region 1050–100 cm^{-1}

In Table 6.2 the Raman bands observed for each specimen are collected and listed for the two wavenumber ranges: namely, A, between 1850 and 1000 cm^{-1}, and B, between 1050 and 100 cm^{-1}. From these data, several comparisons can now be undertaken.

For example, from Figs. 6.13a, b and 6.14a, b we can ascertain several Raman bands which can be assigned to mineral features in the Nantgarw porcelain body:

- In the higher wavenumber shift region between 1850 and 1000 cm^{-1}, several broad and strong bands occur which are characteristic of laser-excited electronic transitions from rare earth lanthanide complexes which are found naturally occurring in very small amounts approximating to only several micrograms per gm in silicates, which comprise the bulk of fired porcelain bodies. These have been reported previously in the literature by Widjaja et al. (2011) who studied ancient Yuan, Ming and Qing Chinese hard paste porcelain shards, and a recent paper by Carter et al. (*Heritage Science*, 2017) discusses them in some detail: the Raman spectra in this latter paper were derived from Ming period porcelains recovered from the wrecks

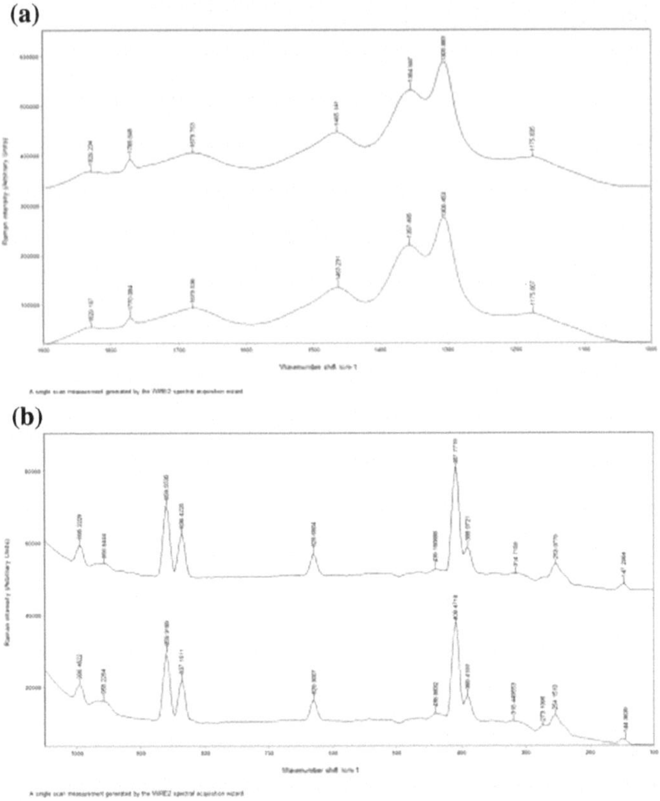

Fig. 6.14 Raman spectral stackplot of Nantgarw crested armorial dinner plate (upper spectrum) and locally decorated dessert plate (lower spectrum): **a** wavenumber region 1850–1000 cm^{-1}; **b** wavenumber region 1050–100 cm^{-1}

of Portuguese carracks which foundered off the Cape of Good Hope in the early 17th Century. In a Ming Dynasty shard (Fig. 6.15) from the wreck of the *Santa Maria Madre de Deus*, Carter et al. (*Heritage Science*, 2017) observed the luminescence bands in the high wavenumber region occurring at 1315, 1348, 1414, 1474, 1609, 1784 cm^{-1}, which show some similarity to those identified here for English and Welsh porcelains. Although strictly belonging to the category of hard paste porcelains and fired at higher temperatures than their European soft paste porcelain counterparts, the silicate matrices in hard paste porcelains will have some structural commonalities and lanthanide element impurities which cause the excitation of these electronic spectra by 785 nm laser excitation in the near infrared region of the electromagnetic spectrum. Hence, the bands which have peak wavenumbers of 1860 w, 1832 mw, 1679 mw, 1464 ms, 1356 s, 1306 vs, 1176 w, and 1145 vw, as will be seen in Table 6.2, can all be assigned to the lanthanide electronic transitions in their host three-dimensional silicate matrix composed of –O–Si–O– and

Table 6.2 Raman spectral bands for porcelains studied

Factory	Item/specimen no.	Bands/cm^{-1}							
(A) Wavenumber region 1850–1000 cm^{-1}									
Nantgarw	Unglazed shard/1	1830	1770	1680	1463	1357	1307	1178	
Nantgarw	Glazed shard/3	1831	1769	1678	1462	1357	1306	1175	
Nantgarw	Cyl. spill vase/6	1829	1770	1680	1465	1359	1307	1200	1177
Nantgarw	Saucer/10	1831	1770	1680	1464	1356	1306	1200	1177
Nantgarw	Armorial plate/24	1829	1770	1679	1465	1355	1306	1200	1176
Nantgarw	Local plate/26	1829	1770	1679	1464	1357	1306	1200	1176
Swansea	Soup dish/9	1830	1770	1681	1520	1465	1344	1305	1189
Swansea	Violeteer/17	1829	1769	1682	1520	1455	1345	1304	1186
Swansea	Watering can/18	1827	1769	1682	1520	1465	1344	1304	1188
Swansea	Trumpet spill vase/7	1826	1770	1681	1520	1462	1343	1304	1188
Swansea	Glassy tea bowl/8	1826	1770	1681	1520	1462	1345	1304	1190
Swansea	Trident plate/15	1826	1770	1680	1520	1462	1345	1304	1190
Swansea	Tea cup and saucer/14	1826	1770	1681	1520	1462	1343	1304	1188
Derby	Cyl. Spill vase/28	1832	1770	1680	1486	1351	1308	1108	
Derby	Barry-Barry plate/25	1825	1770	1680	1467	1351	1308	1108	
Derby	Slop bowl/16	1826	1770	1680	1464	1352	1306	1178	
Derby	Custard cup/29	1828	1770	1678	1465	1357	1306	1175	
Coalport	Sucrier/19	1831	1770	1681	1520	1463	1348	1305	1189
Rockingham	Trumpet spill vase/20	1833	1770	1682	1520	1465	1348	1305	1184
Pinxton	Tea cup and saucer/22	1829	1770	1680	1469	1359	1308	1200	1175
Worcester	BFB coffee can/21	1810	1771	1514	1397	1305	1194		
?	Platter/11	1829	1770	1682	1520	1465	1347	1307	1192
?	Trumpet spill vase/13	1771	1600	1502	1325	1194			
?	Coffee cup/12								
?	Cyl. spill vase/23	1832	1770	1680	1520	1465	1344	1304	1184
?	Slop bowl/27	1828	1770	1675	1463	1355	1306	1174	
(B) Wavenumber region 1050–100 cm^{-1}									
Nantgarw	Unglazed shard/1	997	968	957	912	860	837	631	435
Nantgarw	Glazed shard/3	997	965	957	910	860	836	630	435
Nantgarw	Cyl. spill vase/6	997	967	957	912	861	838	631	435
Nantgarw	Saucer/10	997	968	959	912	860	836	630	435
Nantgarw	Armorial plate/24	995	966	957	927	912	860	836	630
Nantgarw	Local plate/26	996	966	958	927	912	860	837	630

(continued)

Table 6.2 (continued)

(B) Wavenumber region 1050–100 cm^{-1}									
Swansea	Soup dish/9	997	968	957	860	837	631	439	409
Swansea	Violeteer/17	995	968	957	859	836	630	438	408
Swansea	Watering can/18	992	968	955	859	836	630	435	407
Swansea	Trumpet spill vase/7	992	968	958	860	836	631	435	409
Swansea	Glassy tea bowl/8	995	968	958	860	837	631	435	409
Swansea	Trident plate/15	997	861	837	631	409	390	318	270
Swansea	Tea cup and saucer/14	992	968	955	860	837	631	435	409
Derby	Cyl. Spill vase/28	998	968	956	860	836	630	510	470
Derby	Barry-Barry plate/25	994	960	958	860	836	630	510	439
Derby	Slop bowl/16	990	966	960	859	834	438	408	389
Derby	Custard cup/29	990	965	958	860	835	630	436	408
Coalport	Sucrier/19	994	955	859	836	631	407	387	270
Rockingham	Trumpet spill vase/20	996	967	957	860	837	630	440	409
Pinxton	Tea cup and saucer/22	996	967	958	860	837	631	543	431
Worcester	BFB coffee can/21	996	861	838	629	430	409	390	318
?	Platter / 11	996	965	958	861	839	631	439	409
?	Trumpet spill vase/13	996	861	838	631	430	409	390	317
?	Coffee cup/12								
?	Cyl. spill vase/23	997	970	960	860	837	628	430	409
?	Slop bowl/27	990	970	960	859	836	629	438	408
(B) Wavenumber region 1050–100 cm^{-1}									
Nantgarw	Unglazed shard/1	409	389	272	254	245			
Nantgarw	Glazed shard/3	408	389	273	250	245			
Nantgarw	Cyl. spill vase/6	410	390	273	254	245			
Nantgarw	Saucer/10	409	388	270	254	248			
Nantgarw	Armorial plate/24	431	408	389	315	273	253	245	147
Nantgarw	Local plate/26	438	408	389	315	273	254	245	145
Swansea	Soup dish/9	390	317	271	255	149			
Swansea	Violeteer/17	387	316	271	253	145	119		
Swansea	Watering can/18	389	315	270	253	148	119		
Swansea	Trumpet spill vase/7	387	318	270	253	149			
Swansea	Glassy tea bowl/8	389	318	270	253	149			
Swansea	Trident plate/15	253	149						
Swansea	Tea cup and saucer/14	390	318	270	253	149			

(continued)

Table 6.2 (continued)

(B) Wavenumber region 1050–100 cm^{-1}									
Derby	Cyl. Spill vase/28	439	389	318	273	258	122		
Derby	Barry-Barry plate/25	408	389	318	273	253	148	120	
Derby	Slop bowl/16	320	275	255					
Derby	Custard cup/29	387	319	270	250				
Coalport	Sucrier/19	253	145						
Rockingham	Trumpet spill vase/20	388	320	273	255	147			
Pinxton	Tea cup and saucer/22	409	389	318	270	254	147		
Worcester	BFB coffee can/21	254	240	147					
?	Platter / 11	390	318	270	256	149			
?	Trumpet spill vase/13	272	253	243	147	119			
?	Coffee cup/12								
?	Cyl. spill vase/23	390	320	275	252				
?	Slop bowl/27	390	320	275	260				

>Si=O bridging units, and composed of wollastonite, bytownite and other high temperature stable silicates. Widjaja et al. (2011) have quoted similar bands, but not at identical wavenumbers, in their studies of Chinese porcelains with observed bands at 1145, 1216, 1256, 1320, 1420, 1460, 1510, 1585 and 1710 cm^{-1}. This difference is to be expected as it is extremely unlikely that identical lanthanide elemental compound impurities would be found in the local Chinese kaolin sources compared with the analogous European sourced material. Whereas Widjaja et al. employed a spectral reconstruction technique, called BTEM (band target entropy minimisation—a self-modelling curve resolution technique) to remove the lanthanide luminescence bands from this wavenumber shift region of the Raman spectrum, so exposing underlying broad spectral bands arising from the complex silicates, this is actually not so helpful in determining the molecular compositional data in the lower wavenumber region, which is free from such lanthanide luminescent spectral interference. These authors report that the lanthanide spectra represented one component form six individual components identified in the spectral reconstruction process. Hence, it can be seen below that the bands in the wavenumber shift region of the Raman spectra between 1000 and 100 cm^{-1} are probably going to be most informative for the phosphatic, feldspathic, haematite and other components, form which information can be accessed without resorting to spectral data manipulation, itself subject to potential error introduction particularly for the weaker intensity bands. Earlier mention of these luminescent features and their attribution to lanthanide complexes undergoing resonant enhancement under near-infrared wavelength excitation has been noted (Smith and Dent, *Modern Raman Spectroscopy: A Practical Approach*, 2005; Bowie et al., *Handbook of Vibrational Spectroscopy*, 2002; Colomban et al. 2004; Carter et al. 2012) for

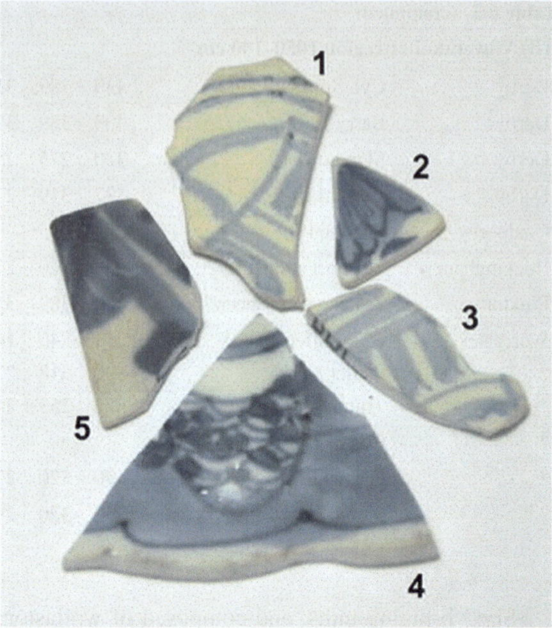

Fig. 6.15 Ming porcelain shards, ca. 1600, recovered from the wrecks of Portuguese carracks attempting to round the Cape of Good Hope from China with cargoes of Chinese hard paste porcelain for the European Market. Raman spectroscopic analysis of the porcelain body reveals the presence of the mineral anatase, titanium (IV) oxide, which occurs naturally along with kaolin, china clay. (Ref: E.A. Carter et al. 2017)

some porcelains and obsidians. Widjaja et al. (2011) analysed the Raman spectra of 3 Yuan Dynasty, 2 Ming Dynasty and 1 Qing Dynasty shards, using 785 nm excitation, and the luminescence originated only in the Yuan and Ming shards and was absent from the Qing Dynasty shard, the spectral bands being very similar in wavenumber for the Yuan and Ming Dynasty shards. In the lower wavenumber shift region, Figs. 6.13b and 6.14b, occur weaker Raman bands which can be assigned to molecular vibrational modes of discrete entities comprising phosphates and silicates: the broad feature centred near 960 cm^{-1} can actually be deconvoluted into two components at approximately 968 and 958 cm^{-1}, which can both be assigned to P–O stretching vibrations in phosphatic components such apatite and whitlockite. This feature is very characteristic of porcelains that have had bone ash, which analyses as predominantly calcium hydroxyapatite, as a component in the paste recipe before firing. Reactions between the phosphate radicals and the silicates at the high temperatures in the kilns result in the formation of the phosphatic minerals described above. The observation of this feature near 960 cm^{-1}, therefore, is convincing spectroscopic analytical evidence for the inclusion of a bone ash component in the porcelain paste and would not be expected to be found for example in hard paste porcelains of Chinese manufacture. An additional feature occurs in this region at 998 cm^{-1}, which will be discussed later.

- Other bands in this low wavenumber region are two rather weak features at approximately 630 and 420 cm^{-1} which can be alternately assigned to the bending vibrations of the $>PO_2$ modes in phosphatic minerals or possibly the Ti=O stretching modes in rutile. The latter mineral occurs in low concentrations in kaolin deposits in Cornwall which were believed to be the source of china clays for both Dillwyn's and Billingsley's productions at Nantgarw and Swansea. A further broad band at 438 cm^{-1} can be attributed to rutile or anatase.

- Other shards from specimens 1–5 delivered similar quality spectra from several independently interrogated sampling replicates, demonstrating the effective reproducibility of the spectral data.
- An assignment of the broad featured envelope near 968 cm^{-1} is the occurrence of wollastonite (key bands at 970, 638 and 340 cm^{-1}), which itself provides a powerful support for the presence of a soft paste porcelain as wollastonite is unstable at higher kiln temperatures so is not detected in hard paste porcelains fired at temperatures near 1400 °C. Hence, recent research on Ming Dynasty hard paste porcelains has failed to determine the presence of wollastonite, which has been detected in a number of European porcelains, including early examples of Medici porcelain of the 16th Century (Mancini et al. 2016; Colomban and Treppoz 2001).

6.7 Interpretation of Data Relating to the Components of the Porcelain Bodies

In summary, therefore, it is now appropriate to consider the important raw material components that were involved in the manufacture of porcelains at Nantgarw and Swansea and also at several related competitor factories where relevant in the early 19th Century: this information has been gathered in Table 6.3.

Figure 6.16a, b show a comparison stackplot of Nantgarw and Swansea porcelain Raman spectra, the former from the Nantgarw spill vase shown in Fig. 1.1 (specimen 6) and the latter from the Swansea duck-egg soup dish (specimen 9) shown in Fig. 1.2. The purpose of this stackplot figure is to demonstrate, firstly, that the Raman spectra of Welsh porcelains are not too dissimilar on cursory inspection and that similar quality spectra can be achieved from perfect specimens of real porcelain which are comparable with those observed from the five Nantgarw shards discussed earlier. In Fig. 6.16a it can be observed that, as expected, the electronic bands in the 2000–1000 cm^{-1} region apparently look very similar, but closer inspection reveals that the Swansea bands are rather broader in the envelope and this can be ascribed to the addition of ground flint glass cullet to the paste frit, which increases the silicate structures underlying the electronic excitation bands in this wavenumber region. In particular, there is a new very broad feature centred at 1510 cm^{-1} which can be assigned to novel silicon-oxygen units from this source. This is absent from the Nantgarw silicate band envelope spectrum because the Nantgarw porcelain recipe did not have glass cullet as an additive. This can therefore be used as a potential diagnostic indicator for the analytical distinction between Swansea and Nantgarw porcelains in this region. Secondly, the Swansea porcelain spectrum shows a weaker but spectrally

Table 6.3 Porcelain formulation components

Factory	Component					
	Kaolin	Chinastone	Bone ash	Glass frit	Sand	Potash
Nantgarw	*		*		*	*
Swansea: Duck-egg	*		*	*	*	*
Swansea: Trident	*	*	*	*	*	*
Swansea: Glassy	*		*	*	*	*
Derby	*	*	*	*	*	
Rockingham	*	*	*	*	*	
Coalport	*	*	*	*	*	*
Pinxton	*	*	*	*	*	
Worcester BFB		*	*	*	*	
Alternative descriptors for components:	China clay	Petuntse	Calcined bone	Cullet	Quartz	Alkaline flux
		Soaprock				
		Cornish stone				

Other minor component additives: arsenic oxide, smalt, pearl ash, borax, calcite, alabaster, gypsum

significant band at 1770 cm^{-1}, which can be assigned to the first overtone of the strong bands occurring near 870 cm^{-1} in spectrum region (Fig. 6.15b).

The lower wavenumber region of the two spectra of the Nantgarw spill vase and the Swansea deep soup dish in Fig. 6.15b shows the following differences:

- The broad phosphatic band centred near 960 cm^{-1} has a different envelope in Swansea porcelain compared with Nantgarw which reflects a significant change in the constituent components: whereas the Nantgarw phosphatic band envelope in this region shows two peaks of equal intensity at 964 and 958 cm^{-1}, which has been interpreted as representing two major phosphate components, probably whitlockite and apatite, that of the Swansea analogue has a major intensity component at 959 cm^{-1} and a minor, weaker intensity component centred around 965 cm^{-1}.
- The creation of newer phosphatic features is reflected in the bending modes spectrum bands near 631 and 409 cm^{-1}, ascribed to slightly different phosphate structures in the porcelain.
- The most significant difference between Swansea and Nantgarw porcelains as provided by their Raman spectra in Fig. 6.16b is seen in the doublet at 861 and 836 cm^{-1}: this is characteristic of a forsterite silicate structural moiety, and can be assigned to a silicate of general formulation Mg_2SiO_4, but combination with iron

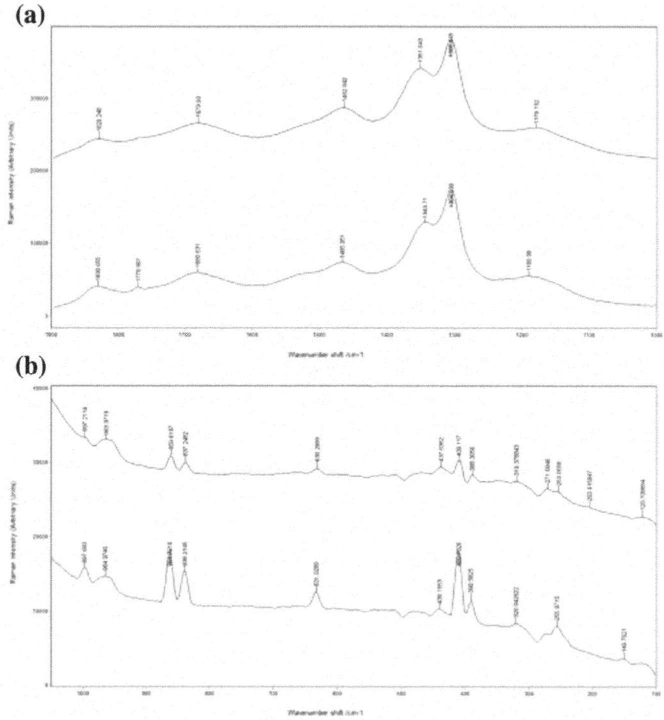

Fig. 6.16 Raman spectral stackplot of Nantgarw cylindrical spill vase (upper spectrum) and Swansea deep soup dish (lower spectrum): **a** wavenumber region 1850–1000 cm^{-1}; **b** wavenumber region 1050–100 cm^{-1}

oxide renders a $(Mg,Fe)SiO_4$ structure which gives a relative change in intensity and band wavenumber of each component of the 860/830 cm^{-1} doublet: we would anticipate the formation of such a species by addition of a magnesium-rich component to the paste. Forsterite, Mg_2SiO_4, and fayalite, Fe_2SiO_4, are the two extreme members of the olivine series of minerals which have characteristic Raman signatures arising from the symmetric and asymmetric Si–O stretching in SiO_4 orthosilicate occurring at wavenumbers of 820 s and 850 ms in fayalite and with reversed relative intensities at wavenumbers of 835 ms and 860 s in forsterite. Previous elegant Raman spectroscopic studies of fayalite and forsterite mixtures and of the fayalite-forsterite solid solution (Foster et al. 1993) have revealed the presence of progressive changes in wavenumbers and intensities which occur between the two extreme members of this olivine series. In our porcelain spectra recorded here, the presence of the characteristic doublet at wavenumbers of 860 and 830 cm^{-1} firstly is indicative of an olivine, and secondly the relative intensities of the doublet band and their wavenumbers at or near 860 and 830 cm^{-1} confirms that the species form which these bands originate is compositionally very close to forsterite, with a formulation of Mg_2SiO_4, a magnesium-rich orthosilicate formed from steatite at high temperatures. Both fayalite and forsterite are stable at high temperatures in excess of 2000 °C and they find use in the manufacture of refractory materials. This is certainly strongly indicative of Dillwyn's addition of steatitic minerals to his Swansea porcelain paste and is compositionally not verifiably present to a significant extent in the Nantgarw analogue as seen in Fig. 6.16b. It is of passing interest to note that forsterite in mineral form is a beautiful green in colour and certainly would not detract from the duck-egg translucency of the incipient Swansea porcelain! However, it should not be theorised from this that the duck-egg colour of Swansea porcelain arises from the presence of a mineral olivine component from steatite addition as in that case we should expect many more porcelains to exhibit this quality. Dillwyn, an accomplished chemist and Fellow of the Royal Society, was aware of the light transmission properties of minerals and he is recorded in making a purchase of cobalt blue, a cobalt aluminate, or of smalt, a cobalt aluminosilicate, for addition to his porcelain paste to offset any yellow cast caused by oxides of iron in impure sand. Clearly, therefore, spectroscopically the presence of the strong forsterite doublet and of flint glass are discriminators between Swansea and Nantgarw porcelains, the former possessing the spectral signatures but the latter does not. As expected for this hypothesis, we should expect to see an increase in the presence of forsterite in the trident Swansea porcelain body, and this will be confirmed later—and, of course, trident porcelain has lost the duck-egg translucency of the finest Swansea body, so confirming that steatite addition is not responsible for the attractive and desirable duck-egg, blue-green translucency of the deep Swansea soup dish considered here and seen in Fig. 1.3.

- It can be concluded, therefore, that Raman spectroscopy thus far offers a hitherto unrecognised discriminatory analytical facility for identifying Swansea and Nantgarw porcelains nondestructively without interference from the glaze or enamelled surface features on perfect finished porcelain articles.

6.8 Discrimination Between the Swansea Bodies

Several Nantgarw pieces have been studied in this sampling exercise, namely the shards numbered 1–5, a spill vase number 6, a coffee cup and saucer number 10, armorial crested dinner plate number 15 and a locally decorated dessert plate numbered 16. All have identical Raman spectra, indicating that they have the same paste body composition, affirming within the range of exemplars studied that there was only ever one composition body manufactured at Nantgarw. However, in our specimen set there are several Swansea pieces which clearly have different attributed body compositions (a duck-egg porcelain spill vase, specimen number 7, a soup dish specimen number 9, and a tea cup and saucer specimen number 11; a glassy porcelain teabowl specimen number 8; a trident porcelain dessert plate specimen number 12) and it will be relevant next to determine if the Raman spectroscopic technique is able to discriminate between these bodies. Figure 6.17 shows a spectral stackplot of these Swansea bodies to verify the ability of the analytical procedure to discriminate

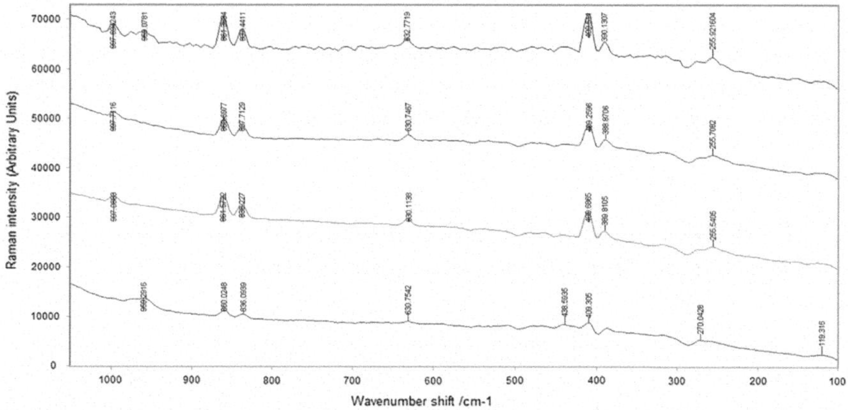

Fig. 6.17 Raman spectra of exemplars of the three identified Swansea porcelain bodies–duck-egg, trident and glassy porcelains. From the top, duck-egg porcelain, saucer from William Billingsley signed Swansea tea cup and saucer (Fig. 6.29); glassy porcelain teabowl (Fig. 6.23); trident porcelain dessert plate decorated by David Evans (Fig. 1.6); duck-egg porcelain soup dish decorated by William Pollard (Fig. 1.2). Wavenumber region (**b**), 1050–100 cm^{-1}

effectively between these three major porcelain types: in this stackplot the spectra represent from the top, the duck-egg saucer, the glassy teabowl, the trident plate and the duck-egg soup dish.

- The duck-egg bodies at the top and bottom are identical and differ from the glassy and trident bodies shown between them. The broad feature centred at about 997 cm^{-1} has a broad counterpart centred at 960 cm^{-1} only in the two duck-egg bodies, which therefore clearly indicates the latter is probably best assignable to the phosphatic component of bone ash, which is present in the duck-egg recipe but not significantly in the other two. In contrast, the higher wavenumber feature is therefore probably due to a silicate feature in a feldspathic component, which is present in all three bodies. The characteristic doublet at 838 and 860 cm^{-1} is present in all three bodies, confirming the assignment of this feature as arising from a magnesium silicate such as forsterite, Mg_2SiO_4. Likewise, bands at 632, 409, 390, 320 and 255 cm^{-1} are present in all four specimens of Swansea porcelain. The 632 cm^{-1} feature is assignable to several possible silicate components, including mullite and muscovite, as well as a possible bending mode of phosphatic apatite. The series of bands at 409 cm^{-1} and below could be Fe–O stretching modes of haematite from the fine river sand used in the recipe.
- The trident and glassy bodies are spectroscopically very similar in high temperature species composition—which is interesting historically since these represent the first and last formulations attempted at the Swansea China Works, from the earliest commercial glassy porcelain marketed in 1814/1815 to the final trident soapstone body made in 1819/1820.

- For ease of presentation at this point the unmarked Swansea spill vase Raman spectra (specimen number 7) have not been reproduced here but, as expected they are identical to those of the duck-egg saucer and the soup dish, confirming its attribution as Swansea. Further details will be discussed below.

6.9 Evaluation of Marked and Unmarked Welsh Porcelains Against the Raman Spectroscopic Protocol

It is clear that from these preliminary, nondestructive analytical Raman spectroscopic studies that a protocol for the evaluation of Nantgarw and Swansea porcelain has emerged and it now remains to test this spectral protocol against further specimens of suspected Welsh porcelain. and finally some porcelain from contemporary factories. This has been accomplished as follows against the listing of specimens given above in Sect. 6.1.

6.9.1 Swansea Spill Vase

A trumpet shaped spill vase (specimen numbered 7) and shown in Fig. 1.9, with beautiful duck-egg translucency (Fig. 1.10) characteristic of the finest Swansea body, unmarked, has been decorated with a dancing figure of a Chinaman in a background of exotic palm trees and foliage, with tendrils and fronds of gilt seaweed in the characteristic manner of William Billingsley. The pattern is unrecorded but the trumpet spill vase is of a rare Swansea shape (Jones and Joseph, *Swansea Porcelain*, 1988). The Raman spectrum of this spill vase is identical with that of the duck-egg porcelain body composition of the William Billingsley marked cup and saucer (specimen numbered 11, Figs. 6.10 and 6.11), see Fig. 6.18, and the William Pollard decorated and marked Swansea soup dish (specimen numbered 9, Figs. 1.2 and 1.3), so confirming its attribution definitely as a Swansea manufacture.

6.9.2 "Swansea" Platter

The specimen shown in Fig. 6.6 is of a meat platter which possesses a fine duck-egg translucency and also a SWANSEA red mark on the reverse (Fig. 6.7): this mark has been observed hitherto on Swansea porcelain and is rather unusual in that it is not a stencilled mark but is composed of hand drawn capital letters followed by a full stop. Jones and Joseph (*Swansea Porcelain*, 1988) have discussed this particular mark, which has been attributed by them to the hand of Henry Morris or alternatively to the workshops of the Swansea china London retailer Apsley Pellatt

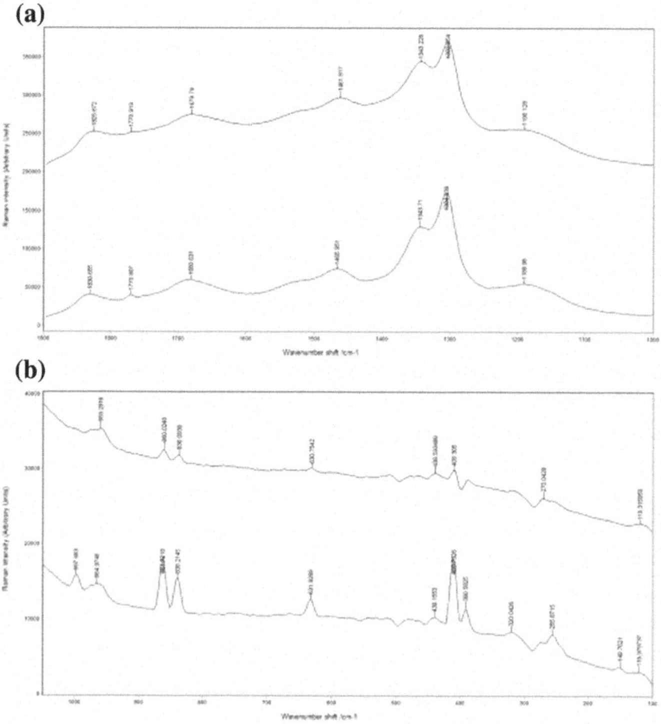

Fig. 6.18 Raman spectral stackplot of Swansea trumpet spill vase (Fig. 1.9) decorated with a chinoiserie scene (upper spectrum) and the William Billingsley signed Swansea saucer (Fig. 6.29) (lower spectrum). Wavenumber region (**b**), 1050–100 cm^{-1}

and Green of St Paul's, where it has been accompanied by their trade mark and name. Secondly, the grouping of flowers has been noted by Jones and Joseph as being characteristic of a Swansea decoration sometimes found on porcelain in a moulded form. A more recent discussion of this particular specimen has been undertaken by Edwards (*Nantgarw and Swansea Porcelains: A Scientific Reappraisal*, 2017a, b, c, d) where another possibility was considered, namely, that this could have been a Staffordshire porcelain item purchased by Morris and decorated by him, and marked by him, in Swansea. Morris is known to have undertaken commissions of this sort and had his own muffle furnace for enamelling and glazing pieces bought in the white during the period 1820–1830 after closure of the Swansea factory in 1820. Clearly, this porcelain item therefore provides a classic example requiring analytical data to better define its origin and that analysis has been accomplished here for the first time. The stackplotted Raman spectra of the SWANSEA marked platter and of the Swansea marked soup dish decorated by William Pollard and shown in Fig. 1.2 are given in Fig. 6.19a, b. In Fig. 6.19a, b it can be seen that both spectra are almost identical in band wavenumbers but there are some differences in the relative intensities of several bands, such as the doublet characteristic of forsterite at 860 and

830 cm^{-1}. It can therefore be concluded that the meat platter could be realistically attributed to a Swansea origin, despite its unusual non-standard mark and unusual shape. However, the matter does not end there because another possibility now must be considered: could this plate have been manufactured elsewhere and sold in the white to Morris at Swansea for decoration? The implication, of course, is that the porcelain body which has been determined to be the finest Swansea duck-egg could well have been reproduced identically outside Swansea—using a very similar recipe to that tabled by Dillwyn for his finest porcelain! Dillwyn's experimental notes and paste composition were a closely guarded secret, known only to himself and his kiln manager, Samuel Walker, and it is perhaps feasible to suggest that espionage would have revealed the intricate details perchance to another interested party. In the previous literature on the Swansea China Works and its products a suggestion has been made (Morton Nance, *The Pottery and Porcelain of Swansea and Nantgarw*, 1942) that upon closure of the factory in 1820 and at the disbursement of the Swansea effects, John Rose of Coalport purchased the moulds and secret recipe for Dillwyn's duck-egg porcelain—but this has never been substantiated, and indeed there has never been any evidence of pieces of porcelain with such specific Swansea characteristics being produced at Coalport. One thing is clear, the porcelain body of this platter is very similar to that of genuine Swansea from the analytical spectral data shown here. Another possibility that requires checking involves the acquisition of the Raman band signatures of pieces from other factories for comparison: this was undertaken as part of this exercise and the conclusions will be discussed later, especially for the intriguing case of the "Swansea" platter.

This is an intriguing result and opens the way for further scientific appraisal of Welsh porcelains which have perhaps been dismissed too readily on a more cursory examination as bogus or fake in the past merely because they have not confirmed to established evidence based on shape, mark or size factors. It has already been pointed out by Edwards (*Swansea and Nantgarw Porcelains: A Scientific Reappraisal*, 2017a, b, c, d) that of the 704 recorded set pattern numbers assumed used by the Swansea factory in its lifetime only 67 have thus far been identified after many years diligent research by Jones and Joseph (*Swansea Porcelain*, 1988) and it could well be thought that there are some hitherto undiscovered factory shapes and sizes as well to be accounted for. At the moment, therefore, we must conclude that the scientific evidence could indicate that the meat platter is a potentially rare piece of true Swansea porcelain—this attribution would of course be confirmed by the discovery of other pieces from the same service which possess more typical Swansea shapes and perhaps also be accompanied by a more standard Swansea script mark.

6.9.3 Nantgarw Spill Vase

Another piece of unusual porcelain which is thought to be possibly Nantgarw in origin is shown in Fig. 6.20: this is a spill vase which departs from the usual Nantgarw shape met in the form of a cylindrical slightly tapered item with or without applied

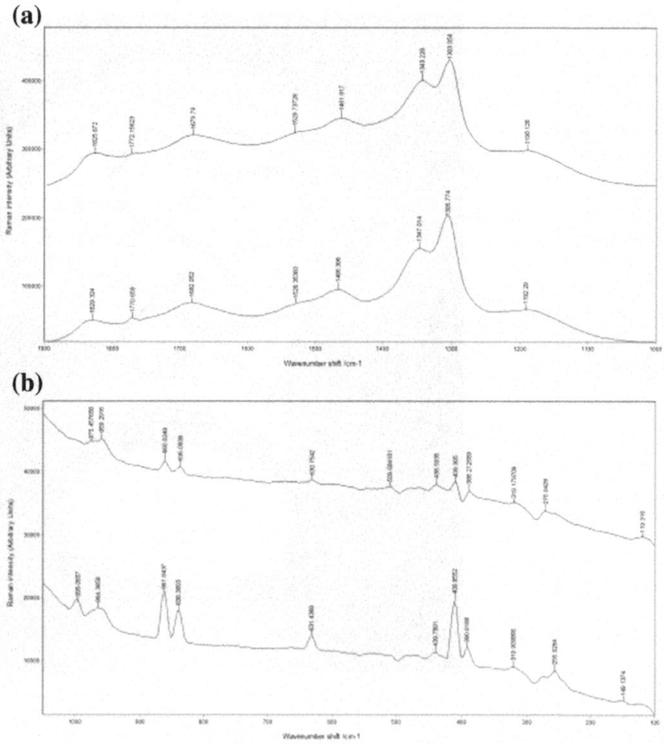

Fig. 6.19 Raman spectra of the possible Swansea meat platter (Fig. 6.6) which has many of the attributes of the finest Swansea duck-egg porcelain paste but is of a previously unrecorded shape for the factory: upper spectrum, platter, marked SWANSEA.; lower spectrum, Swansea deep soup dish with red stencil mark SWANSEA (Fig. 1.2) decorated by William Pollard. **a** wavenumber region 1850–1050 cm^{-1}; **b** wavenumber region 1050–100 cm^{-1}

masks near the top rim. A London decorated spill vase of this usual type is shown in Fig. 1.1 and this has been analysed here along with a Nantgarw shard in Fig. 6.13. The spill vase in Fig. 6.20 is actually a trumpet shape, perhaps more associated in shape with Swansea and with the later versions emanating from the Rockingham factory (Cox and Cox, *Rockingham Porcelain*, 2005). However, the discovery of a trumpet spill vase, badly damaged and unmarked has been reported in the effects of Blickling Hall in Norfolk, which is under the ownership of the National Trust. There the spill vase is catalogued as "*possibly Davenport or Swansea porcelain*": the decorative pattern on this spill vase is identical with that on a Nantgarw tea and coffee service, items from which occasionally appear at auction (a cup and saucer from this set patterned service is shown in Fig. 6.21). The dimensions of the spill vase depicted here are very similar to those of the Blickling Hall example and the translucency is unquestionably Nantgarw as seen in Fig. 6.19. The Raman spectrum of this spill vase is shown in Fig. 6.22, from which it can be concluded that the

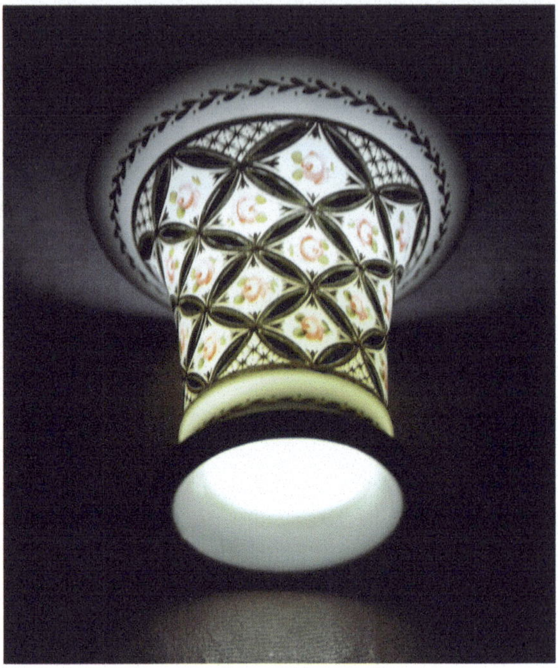

Fig. 6.20 Spill vase, possible Nantgarw porcelain, trumpet shape, of similar dimensions to another (National Trust, Blickling Hall, Norfolk) found to be en suite with a Nantgarw tea service (see Edwards, Swansea and Nantgarw Porcelain: A Scientific Reappraisal, 2017a, b, c, d), photographed in transmitted light to show the superbly clear porcelain body

body analyses as Nantgarw porcelain; however, the decoration is not matched by any known Nantgarw pieces currently identified, so it should now perhaps be formally recorded as a potential Nantgarw example, with an unrecorded pattern awaiting confirmation from the discovery of a marked piece decorated en suite.

6.10 Ornamental Porcelain

One of the most challenging issues for the attribution of porcelain to a particular factory occurs with ornamental pieces, which are usually much rarer than service items or set pieces and most were rarely marked. Their identification therefore is entirely dependent upon expert recognition of a painter's work or even some special idiosyncrasy associated with the factory concerned. Swansea and Nantgarw are not exceptions to this tenet and through history there has been much porcelain accredited or attributed to their manufacture which is not now accepted as genuine. It must be said, however, that a piece of porcelain should not be relegated to being of dubious origin just because it is unique and its shape and style do not match currently marked and accepted versions! Jones and Joseph (*Swansea Porcelain*, 1988) in their comprehensive survey of Swansea set patterns note that only some 67 pattern numbers can now be recognised as "Swansea" from over 700 recorded (i.e. that only some pieces have the pattern number AND a Swansea stencilled or impressed mark)—it

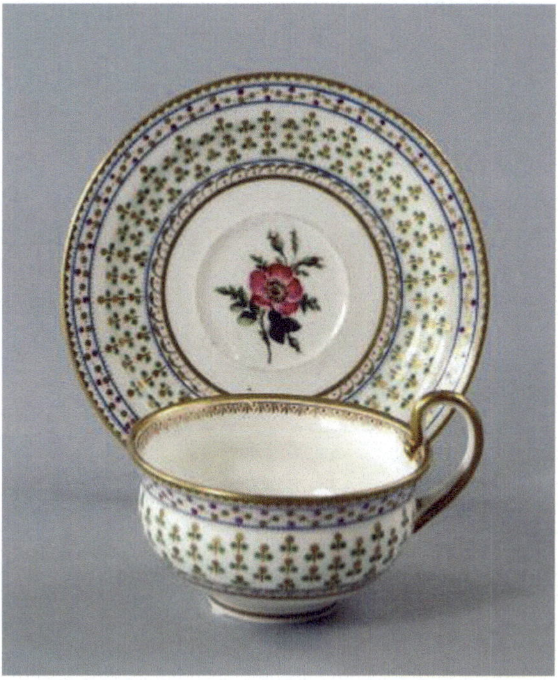

Fig. 6.21 Nantgarw tea cup and saucer with identical pattern to that possessed by the unidentified spill vase in Blickling Hall, referred to in Fig. 6.39

may be inferred, therefore, that 90% of Swansea pattern set pieces could be in circulation but currently unrecognised or designated as unattributable definitively. The answer is clear—the non-destructive analytical interrogation of an unknown piece would reveal whether or not the porcelain is of a Swansea or Nantgarw formulation and this valuable parameter could be input to an informed decision about the true origin of the porcelain in question. The "genuine or fake" label for Welsh porcelain has been has been a focus of discussion for many years since the desirability of owning these works of ceramic art has always been high, especially following the closure of the factories in the first few years of the decade of 1820–1830. Later in the 19th Century, the Samson et Cie factory in Paris set up a business in copying porcelains from famous British and continental European factories, including Sevres, Swansea and Nantgarw: generally, these "copies" reflected the contemporary desires of society which favoured overactive embellishment of the Georgian simplicity of the Welsh porcelains, but the painting was cleverly executed in the appropriate styles of the Swansea and Nantgarw output. A major problem appears to be that the Samson mark, crossed swords in blue overglaze pigment resembling the Sevres factory mark, was easily removed by acid etching—allowing the substitution of another factory mark in its place. Samson et Cie. of Paris was established in 1845 when Edme Samson received a commission to make replacements for important dinner services which had suffered breakages: he expanded the business to cover European and Chinese porcelains until the firm closed in the 1920s. Whilst Samson

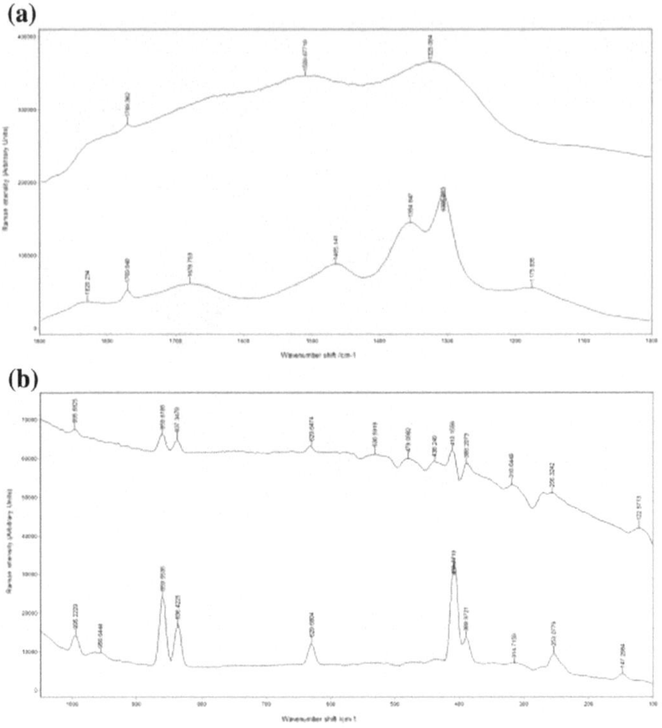

Fig. 6.22 Raman spectra of rare, possibly Nantgarw, trumpet shape spill vase shown in Fig. 6.39 (upper spectrum) and Nantgarw cylindrical spill vase (Fig. 1.1) (lower spectrum). **a** wavenumber region 1850–1050 cm^{-1}; **b** wavenumber region 1050–100 cm^{-1}

claimed that his copies were innocent and not intended to deceive, being marked with script *SS* or crossed swords after the Meissen mark, or with a "red chop" mark on Oriental porcelains, these marks could be easily removed by unscrupulous forgers (Craddock, *Scientific Investigation of Copies, Fakes and Forgeries*, 2009). Although most experts are of the opinion that Samson copies of soft paste porcelains such as Swansea and Nantgarw were easily discriminated against because they were of French hard paste porcelain construction, their copies of Chinese or Meissen wares are much more difficult to identify—stressing the importance of analysis in support of a forensic art investigation. It is hoped, therefore, that non-destructive spectral analysis will do much to redress this imbalance and enable unusual pieces to be examined more closely when otherwise they might be discarded or relegated to the category of "unknown" or "possibly fake". Samson was clearly not the sole copier in this respect and it is nevertheless revealing to learn that Henry Sandon (Sandon, *Starting to Collect Antique Porcelain*, 1997) estimated that some 16% of porcelain he has observed for sale in many country antique auction sales in the UK were copies emanating from Chinese factories which are currently still actively producing such items, which can be classified as copies of both both Oriental and Occidental factory porcelains. Occasionally, however, Samson copies are still found in several eminent collections of Welsh porcelain, where they have been accepted as genuine examples by association.

Even experts such as Morton Nance (*The Pottery and Porcelain of Swansea and Nantgarw*, 1942) who spent over 40 years studying Welsh porcelain in depth admitted he had been taken in by some very clever forgeries, which he donated to the National Museum of Wales in Cardiff for the tuition and instruction of others on the basis of "*caveat emptor*". One example he has cited was a beautifully decorated attributed Nantgarw spill vase with a script "Nantgarw" mark on the underside, which was believed originally to be genuine but is now considered clearly to be a fake! Another unusual piece of "Welsh" porcelain has appeared at auction recently with a red script "Swansea" mark accompanied by another stencilled red enamelled legend, "*Made in England*": here the deficient geographical knowledge of the faker is evident, but the item was still catalogued and sold as Welsh porcelain!

6.10.1 Swansea Violeteer and Watering Can

As a first riposte in the analytical spectroscopic assessment of unmarked Swansea and Nantgarw porcelains which could be considered as being of an ornamental variety (although perhaps not strictly apposite in the case of the violeteer, which does have a functional use), the Raman spectra of a violeteer and watering-can are stack-plotted in Fig. 6.23a, b. Again, these are observed to be identical and therefore we have an independent analytical confirmation of their authenticity as being of Swansea origin, as originally suspected: although in this particular case there is much evidential information in the literature from their shape and decoration that they are the genuine articles, nevertheless, their rarity and resultant premium value would have perhaps made them targets for potential forgery—since it has been estimated that fewer than about six genuine examples of each type are currently known in museums and ceramics collections. The Swansea violeteer (Fig. 6.24), unmarked, duck-egg porcelain, beautifully decorated with bouquets of garden flowers by Henry Morris, with strap handles and a pierced lid with oval holes for containing scented *pot pourri*, can be dated to *ca*. 1815–1819, and is illustrated in John (*Swansea Porcelain*, 1958) and in Jones and Joseph (*Swansea Porcelain*, 1988). The Swansea watering can (Fig. 6.25), unmarked, configured from a coffee can base with the addition of a lip. spout and handle, decorated by William Pollard, ca. 1815–1819, is also illustrated in John (*Swansea Porcelain*, 1958) and in Jones and Joseph (*Swansea Porcelain*, 1988). The Raman spectral stackplot in Fig. 6.23a, b for violeteer and watering-can are identical and both match that of the other fine duck-egg porcelain specimens studied here, such as the deep soup dish (Fig. 1.2) and the basket weave cartouche moulding modelled cup and saucer (Fig. 6.10), both marked with Swansea stencilled or script marks in red enamels (Figs. 1.3 and 6.11). It can therefore be concluded that

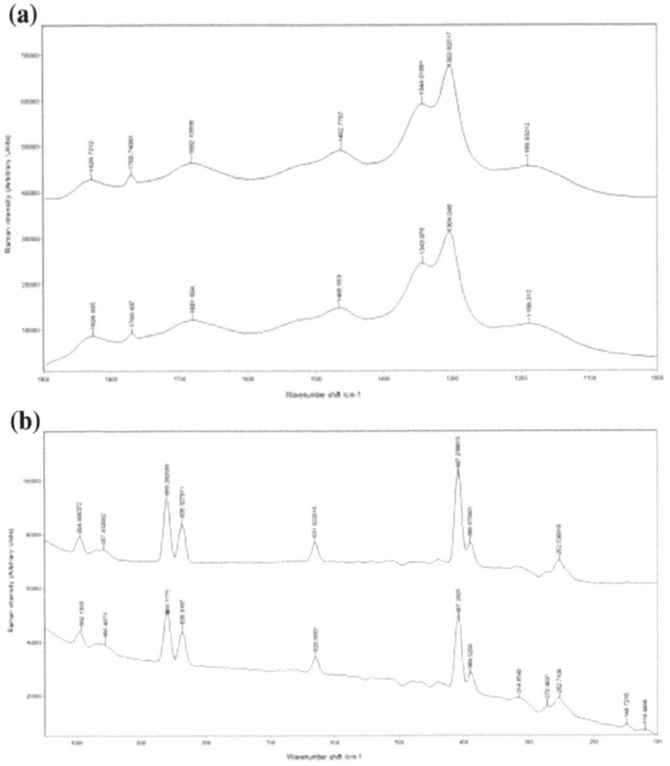

Fig. 6.23 Raman spectral stackplot of Swansea violeteer (Fig. 6.24), decorated by Henry Morris, and a watering can (Fig. 6.25) decorated by David Evans, both unmarked, finest duck-egg porcelain, ca. 1817–1819. **a** wavenumber region 1850–1050 cm^{-1}; **b** wavenumber region 1050–100 cm^{-1}

spectroscopic analysis has unambiguously confirmed the attribution of the violeteer and watering-can to a Swansea origin of manufacture.

6.10.2 London-Shape Coffee Cup

The coffee cup illustrated in Fig. 6.8 exhibits many visible features of a Swansea porcelain attribution, such as duck-egg translucency (Fig. 6.9, compare with Fig. 1.3), dimensions as exemplified by Jones & Joseph (*Swansea Porcelain*, 1988) and the evidence of the characteristic kick-spur triple ogee handle which prominently almost touches the rim of the cup unlike the spur on other contemporary factories. The pattern, which is mainly comprised of very finely gilded flowers and seeds on a pink edge ground colour is not recorded in Jones and Joseph (*Swansea Porcelain*, 1988) and additionally a pattern number "877" under the footrim is higher than any

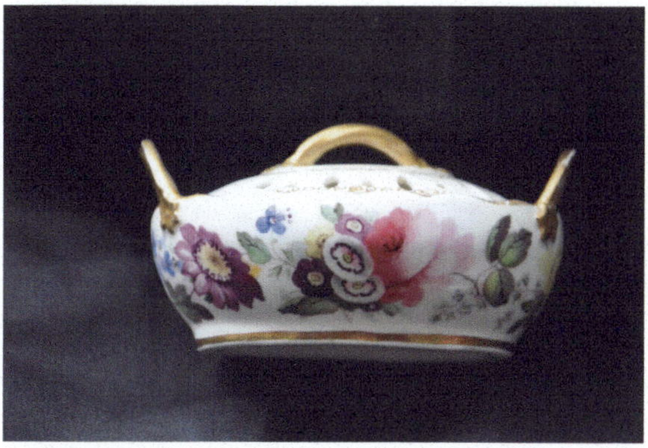

Fig. 6.24 Swansea porcelain violeteer, duck-egg body, with pierced lid, gilt strap handles and Greek scroll gilding border, decorated with sprays of garden flowers by Henry Morris. Unmarked

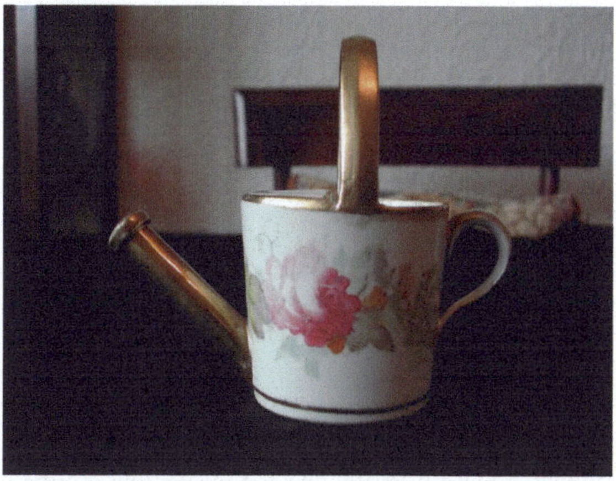

Fig. 6.25 Miniature Swansea watering can, duck-egg body, decorated by David Evans with a wreath of garden flowers; based on a coffee can with an added gilt spout, part cover to the front rim and a strap handle. Unmarked

hitherto recorded for Swansea tea and coffee services. This could be a stumbling block to the definitive attribution of this coffee cup to a Swansea origin. Unfortunately, the Swansea pattern books, if they ever existed, have not survived so historians have no idea exactly how far the pattern numbering sequence ran – the highest recorded pattern identified thus far is 704 (Edwards, *Swansea and Nantgarw Porcelains: A Scientific Reappraisal*, 2017a, b, c, d), but, of course this does not imply that the pattern numbering stopped there and many unnumbered Swansea patterns have still to be identified, as less than 10% of those recorded so far have been matched with numbers. It is perfectly possible, therefore, that this coffee cup is of a Swansea origin. This specimen, therefore, provides a good test example for our spectroscopic

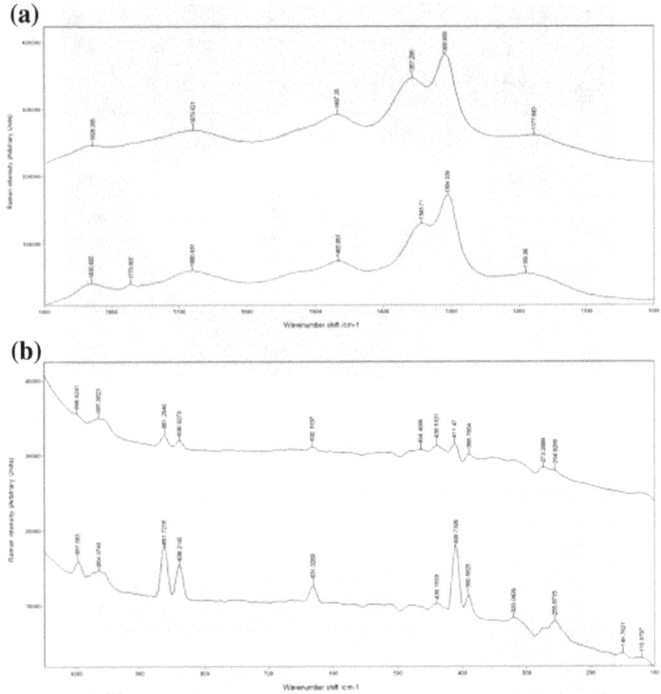

Fig. 6.26 Raman spectral stackplot of potential Swansea coffee cup, unmarked, Fig. 6.27) (upper spectrum) and Swansea deep soup dish (Fig. 1.2) (lower spectrum): **a** wavenumber region 1850–1000 cm^{-1}; **b** wavenumber region 1050–100 cm^{-1}

protocols by which the identification of a Swansea origin or otherwise could be achieved. The Raman spectrum of the suspect coffee cup shown in Fig. 6.8 and compared with that of the Swansea soup dish is given in Fig. 6.26.

A comparison between the Raman spectrum of the coffee cup (Fig. 6.8) and that of the marked Swansea soup dish (Fig. 1.2) indicates that the spectra of the two pieces are very similar indeed: all the major features identified as characterising the Swansea duck-egg porcelain body are present in the spectrum of the coffee cup, with the proviso that whereas there seems to be a similar amount of the phosphatic component attributed to bone ash, there is rather more of the feldspar component in the coffee cup and also possibly the haematite from the lower wavenumber signatures. The latter can be attributed to variability of sourcing the fine river sand needed for the paste mixture and the former could reflect a small experimental variation in feldspathic content, which we know was carried out by Lewis Dillwyn during the manufacture of his duck-egg body in attempts to create a more robust porcelain without loss of fineness, as was eventually evident in the trident wares he produced latterly (Hillis 2005). It can be concluded therefore that the coffee cup could well be of Swansea origin but awaits confirmation that its pattern number is truly assignable to

Swansea set pieces: a further point to note is that if this is a Swansea product, as seems to be supported scientifically and spectroscopically, then it would perhaps be one of the last of the true duck-egg bodies actually to leave the factory from its high pattern number—again supporting the hypothesis that at this stage of manufacture, Lewis Dillwyn was already contemplating changes to his successful duck-egg porcelain composition recipe to reflect his impending movement onto a trident porcelain body, where a soapstone or steatitic component would be used to replace the bone ash. *Verdict*: probably Swansea, but awaiting verification of pattern and pattern number from associated named Swansea services.

6.11 Blemishes in Nantgarw Porcelain

Attention has already been drawn to the appallingly high losses encountered in the firing of Nantgarw porcelain which accounted for up to 90% kiln losses through sagging and warping of the shapes. Prized for its fabulously high translucency as seen in Fig. 1.1, a rather unexpected blemish is sometimes seen on early porcelains and is shown here in Fig. 4.4, which is the reverse side of a locally decorated dessert plate shown in Fig. 4.3. This Nantgarw dessert plate, locally and simply decorated with sprays of orange roses, foliage and blue delphiniums and a simple edge gilding is marked with an impressed NANT-GARW C.W. The translucency is excellent but the presence of black spots and blemishes it is reasoned would have rendered this piece unacceptable for decoration in the London market through the agent Messrs. Mortlocks. Sometimes, small, dark blemishes or pits in an otherwise perfectly acceptable piece of porcelain could be masked by the strategic placement of insects such as butterflies and moths or by small flower buds: an example of this is seen in Fig. 4.2, where several mosquitoes have been used to cover small blemishes the underside rim of a large Nantgarw meat platter from the *Farnley Hall* service (Edwards, *Swansea and Nantgarw Porcelains: A Scientific Reappraisal*, 2017a, b, c, d). A Raman spectroscopic interrogation of some of these dark blemishes on the locally decorated Nantgarw dessert plate has produced the spectrum shown in Fig. 6.27, although several bands of the silicate matrix are still observed at weaker intensity, two new, rather broad, features are seen at approximately 1580 and 1320 cm^{-1} which are characteristic of amorphous carbon, the so-called G (sp^2) and D (sp^3) bands, respectively, of hybridised carbon of graphitic or diamond-like structure. It can be concluded, therefore, that the black blemish on the surface of the Nantgarw dessert plate is elemental carbon.

It is interesting to speculate on the source of these blemishes as shown in Fig. 4.4: the Nantgarw kilns were oxidising in atmosphere (John, *Nantgarw Porcelain*, 1948) and it is perhaps surprising to think that finely divided particulate carbon could survive without conversion to gaseous carbon dioxide. However, inspection of the blemishes on the dessert plate analysed here reveals that they occur on the surface of the biscuit porcelain and have been covered by the glaze. This suggests that, following the removal of the porcelain after first firing and its treatment with liquidised slip for

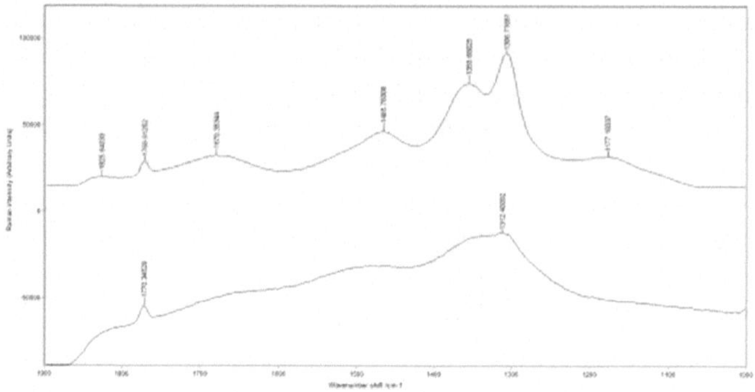

Fig. 6.27 Raman spectrum of black blemish on locally decorated Nantgarw dessert plate (as shown in Figs. 4.3 and 4.4). Wavenumber range 1650–1200 cm^{-1}

glost firing in the glazing kiln, some ingress of carbon particles became trapped in the wet glaze, covered by the glaze and became fixed there upon the second firing. The oxidisation of the entrapped carbon would have been prevented by the hardened surface coating of glaze.

Occasionally, another sort of blemish is seen in Nantgarw porcelain such as that shown in transmitted light in Fig. 6.28, which occurs as a small, cream-coloured and rather diffuse isolated spot seen here near the factory impressed mark NANTGARW C.W. on the armorial crested dinner plate shown in Fig. 6.12. This is entirely different in origin and can be attributed to either a stable mineral impurity in the paste or more probably to imperfect grinding and preparation of the powdered frit—leaving small pockets of feldspar or kaolin. The incorporation of some organic impurities, for example, would certainly give rise to blemishes of this sort and we can look no further than organic contaminants in the bone ash to be responsible for effects of this kind. A Raman spectrum could not be obtained from this blemish, which perhaps was situated too far beneath the surface for effective interrogation to be performed. Both Dillwyn and Billingsley were adamant in their selection of the purest possible sources for their calcined bone ash additive but occasionally some contamination could have permeated through the vetting processes. Such minor and interspersed blemishes do not detract seriously from the generally excellent translucency expected for Nantgarw porcelains and they are certainly of a different order of magnitude for visual impact to those attributed to the particulate carbon discussed earlier. In a contrasting viewpoint, it would also be reasonable to suspect that most organic contaminants would burn off at the high kiln temperatures necessary for porcelain production so maybe the source of this blemish arises from mineral contaminants as suggested.

In another exercise, a similar type of blemish to that found in the Nantgarw dessert plate and assigned to carbon, was found on a Pinxton cup and saucer (Fig. 4.5) dating from 1798, when William Billingsley was producing porcelain for John Coke's enterprise there after his departure from the Derby China Works. This was studied and a small particle of carbon was observed to be encased in a small bubble

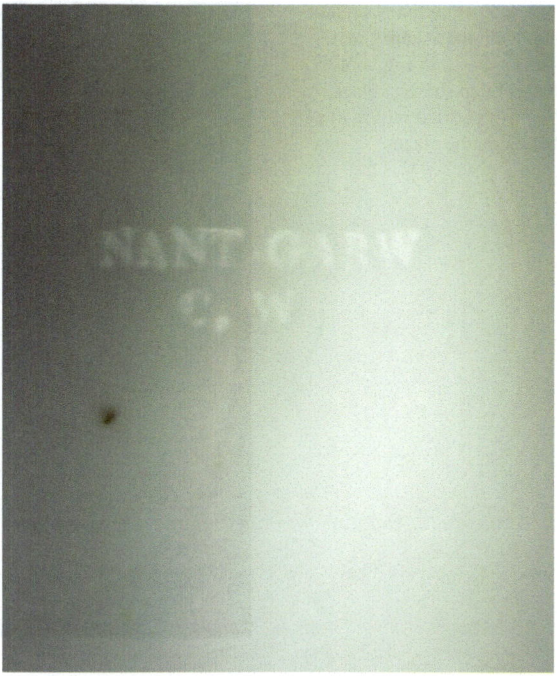

Fig. 6.28 Nantgarw dinner plate, crested armorial decoration at verge, from Fig. 6.31, photographed in transmitted light to illustrate the superb translucency and the impressed underglaze mark, NANT-GARW C.W. Note the very small blemish actually in the porcelain body near the impressed mark

inside the glaze (Fig. 6.29) on the inner surface of the base of the teacup. Here, it could be conjectured that the carbon particle had partially oxidised and the bubble of carbon dioxide produced would then have been trapped in the hardening glaze, so preventing further oxidation from taking place and protecting the carbon particle for further chemical aerial oxidation in the firing kiln. Clearly, William Billingsley did not consider such a blemish sufficiently bad to warrant destruction of an otherwise perfect porcelain piece as he personally decorated the teacup and saucer with some of his finest flower painting.

6.12 Comparator Raman Spectra with Other Factories

As a preliminary exercise in the ability of Raman spectroscopic analysis to discriminate between the Welsh porcelains of Swansea and Nantgarw and others from an approximately contemporary period it was thought useful to examine porcelains from the following factories to see how an emergent protocol for Swansea and Nantgarw would differentiate specimens which might otherwise have been considered in this category. Using the same equipment and under the identical conditions employed for the spectroscopic analysis of the specimens of Swansea and Nantgarw porcelains studied above, the following exemplars were studied:

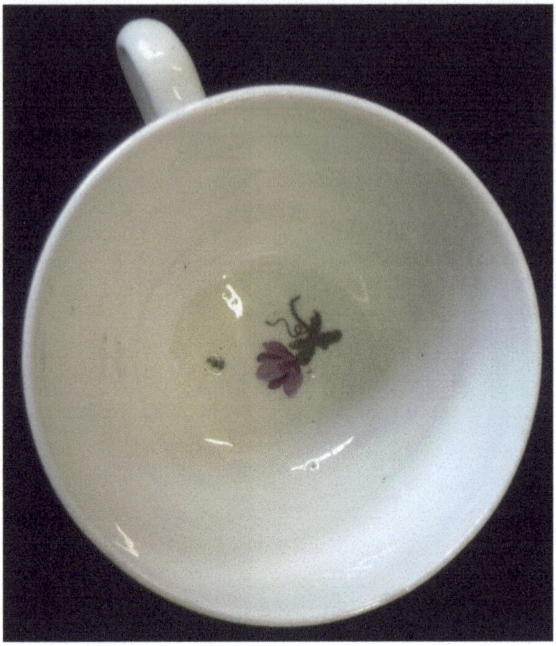

Fig. 6.29 Pinxton porcelain cup from the teacup and saucer shown in Fig. 4.5; base of cup shows a black blemish encased within in a bubble in the glaze

- Coalport: a covered sucrier from a very fine rococo tea service. Pattern number 2/675, circa 1825-30; some of this service is illustrated in Godden (*Coalport and Coalbrookdale Porcelains*, Plate 162, page 263)—where the baroque styling of the bread-and-butter plate, teapot, jug and bowl is typical of the transitional period from Regency simplicity to the full-blown revived rococo characteristic of early Victorian porcelains (Fig. 2.2).
- Pinxton: a very rare tea cup and saucer, decorated by William Billingsley, ca. 1798, from a unique service which is completely ungilded (Gent, *The Patterns and Shapes of the Pinxton China Factory*, 1996, Fig. 4.5).
- Derby: porcelain dinner plate from the Robert Bloor period, ca. 1800–1820, part of the Barry-Barry dinner service (Edwards, *Swansea and Nantgarw Porcelains: A Scientific Reappraisal*, 2017a, b, c, d; Edwards and Denyer, *William Billingsley: The Enigmatic Porcelain Artist, Decorator and Manufacturer*, 2016, Fig. 6.30).
- Worcester, Barr, Flight & Barr period, ca. 1810. Three coffee cans decorated by William Billingsley (Figs. 6.31, 6.32).

Plate from the same service illustrated in John,*William Billingsley*, (1968).

- Unknown Factory: ca. 1815–1820. A beautifully translucent spill vase with Nantgarw-like whiteness to transmitted light and clear white glaze, simply decorated with a young girl in the First Position plie) ballet dancing pose in Regency dress and slippers. Nantgarw shape and dimensions, and a possible attribution except for a red enamelled script mark "908" on the base. As stated above, despite

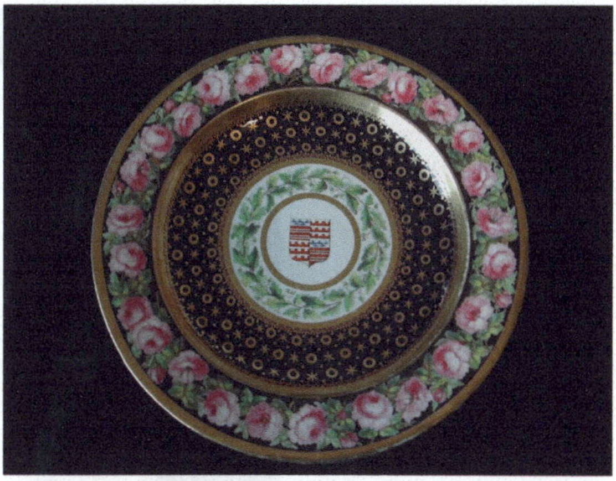

Fig. 6.30 Derby porcelain dinner plate from the Barry-Barry dinner service, ca. 1800–1820, with decoration ascribed to William Billingsley after he had left the Derby China Works. See Edwards (*Swansea and Nantgarw Porcelains: A Scientific Reappraisal*, 2017a, b, c, d; Edwards and Denyer, William Billingsley: The Enigmatic Porcelain Artist, Decorator and Manufacturer, 2016)

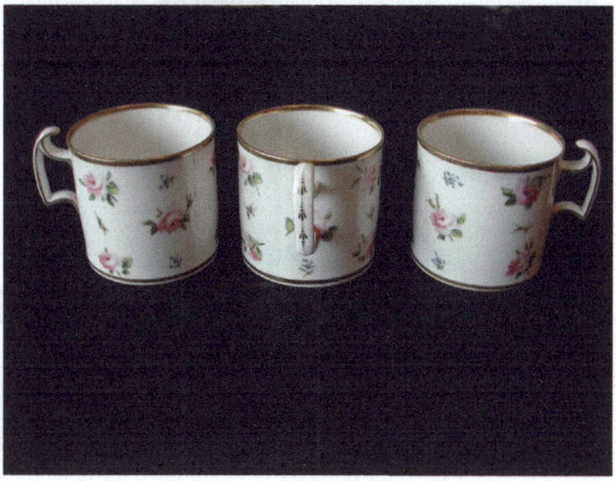

Fig. 6.31 Coffee cans from a Barr, Flight & Barr Worcester porcelain tea and coffee service ascribed to William Billingsley (John, *William Billingsley*, 1968), ca. 1810

its Nantgarw-like translucency, Nantgarw did not mark its pieces with pattern numbers, so this observation must question its possible attribution to the Nantgarw factory. This specimen, therefore, provides a classic example for our nondestructive spectroscopic testing protocols.

Coalport sucrier

Firstly, the Raman spectrum of the Coalport sucrier was stackplotted against that of the putative Swansea platter and the spectra are exhibited in Fig. 6.33a, b. it can be seen that the spectra are very closely similar and it would be realistic to say that they represented the same factory origins. Hence, we could conclude that the platter was of a Coalport origin rather than Swansea as was attested earlier, where some compositional differences were confirmed. This is an interesting result as it appears to confirm that the hypothesis of the platter being made in Coalport and decorated by Henry Morris in Swansea is a correct one: however, the corollary to

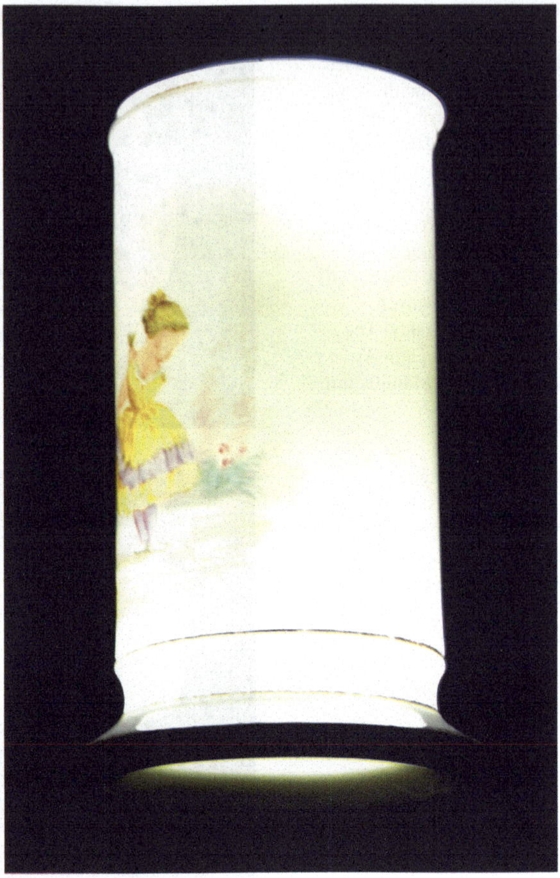

Fig. 6.32 Spill vase, unknown factory, possessing a beautifully clear white translucency similar to that of Nantgarw; unmarked except for a red enamelled number "908" on the underside

this is that the spectra of Swansea duck-egg porcelain and the best Coalport are very closely similar too. Does this imply that John Rose did actually have knowledge of eth Swansea recipe formulation as has been alleged by several authors, following his "purchase" of Swansea items at eth auction sales between 1823 and 1825? It could well mean that he had this real knowledge but this has been smothered in incorrect statements about his acquisition of associated hardware such as the kilns and moulds? The Glynn Vivian Art Gallery in Swansea has an outstanding collection of Swansea porcelain and several items of porcelain from Staffordshire factories bought in by local artists such as Henry Morris, David Evans and William Pollard: one of these is a platter that is similar to that studied here with Morris' decoration—which can now be confirmed as being on Coalport porcelain, which is analytically so very similar to Swansea in paste composition (Fig. 6.33a, b).

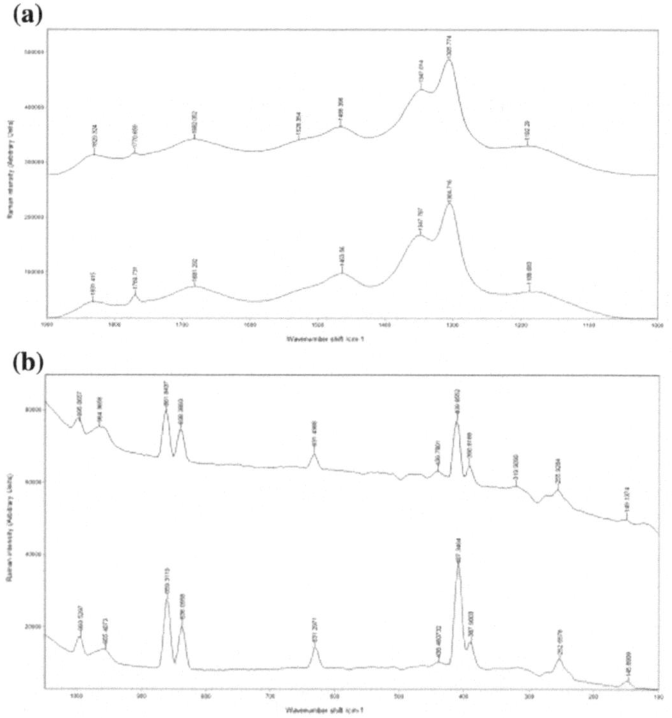

Fig. 6.33 Raman spectra of the Coalport covered sucrier (Fig. 2.2) (upper spectrum). Lower spectrum, the platter marked SWANSEA, which has many of the attributes of the finest Swansea duck-egg porcelain paste but is of a previously unrecorded shape for the factory (Fig. 6.25) and believed decorated by Henry Morris. **a** wavenumber region 1850–1050 cm^{-1}; **b** wavenumber region 1050–100 cm^{-1}

Pinxton cup and saucer/Derby Barry-Barry dinner plate

Secondly, the Raman spectral stackplot in Fig. 6.34a, b provides a comparison between Pinxton and Derby porcelains, exemplified by the William Billingsley cup and saucer and the *Barry-Barry* service dinner plate. The analysis reveals that Billingsley made his own porcelain at Pinxton which had several similarities to that of Swansea in composition and sensible differences with its Derby analogue. This means that analytically it is possible to detect the difference between Pinxton porcelain made at the factory porcelain which was bought in by William Billingsley from Derby for decoration at Pinxton and which was thereafter sold as Pinxton porcelain. This is an important application for the characterisation of pieces which are currently of unknown or uncertain origin regarding the Pinxton factory (Fig. 6.34a, b).

Worcester BFB coffee can

Thirdly, the Raman spectra of a Barr, Flight & Barr Worcester coffee can, shown in Fig. 6.31, decorated by William Billingsley at Worcester in ca. 1810 is shown in

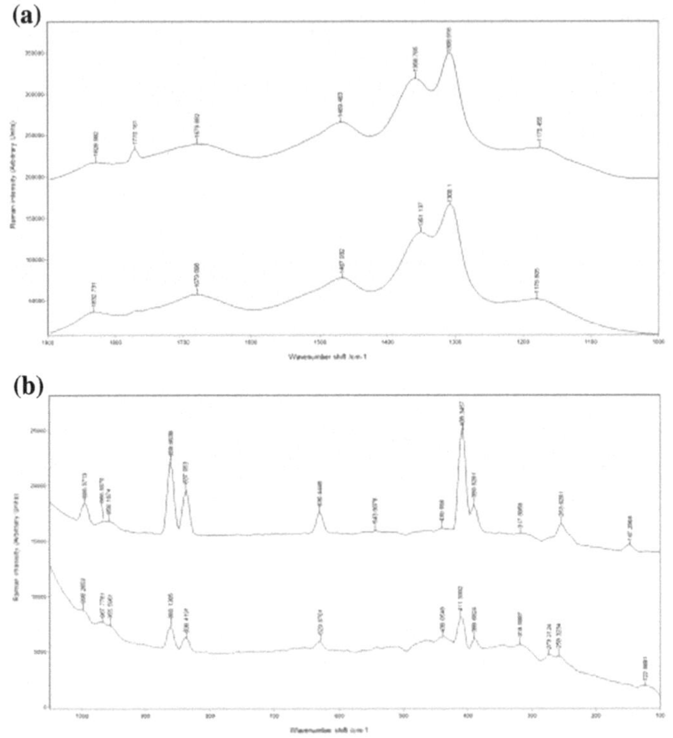

Fig. 6.34 Raman spectra of the Pinxton saucer (Fig. 18), ca. 1798 (upper spectrum). Lower spectrum, the Derby porcelain dinner plate from the Barry-Barry service (Fig. 6.30). **a** wavenumber region 1850–1050 cm^{-1}; **b** wavenumber region 1050–100 cm^{-1}

Fig. 6.35a, b. These spectra are different in many respects to those of the soft paste porcelains of Swansea, Nantgarw, Derby and Coalport as can be seen especially in eth band envelope in Fig. 6.34a. Differences in the phosphatic and feldspar components are reflected in the spectrum in Fig. 6.34b as expected for the Worcester porcelain body, which is best described as a *hybrid* porcelain verging more towards a hard paste classification (Fig. 6.35).

Unknown factory, spill vase

Fourthly, the Raman spectra of the Nantgarw-like spill vase shown in Fig. 6.32 are shown in Fig. 6.36a, b. Although the spectra are in fact very similar to those of the verified Nantgarw specimens studied earlier, the relative compositional differences in the porcelain paste are manifest in the relative intensity differences in the two Raman spectral wavenumber regions shown here. Hence, it can be concluded that the spill vase cannot be attributed to a Nantgarw origin, although the spectra are admittedly reasonably close in similarities. The concern expressed initially about the presence of a pattern number on the spill vase, which would be highly unusual if not unique for

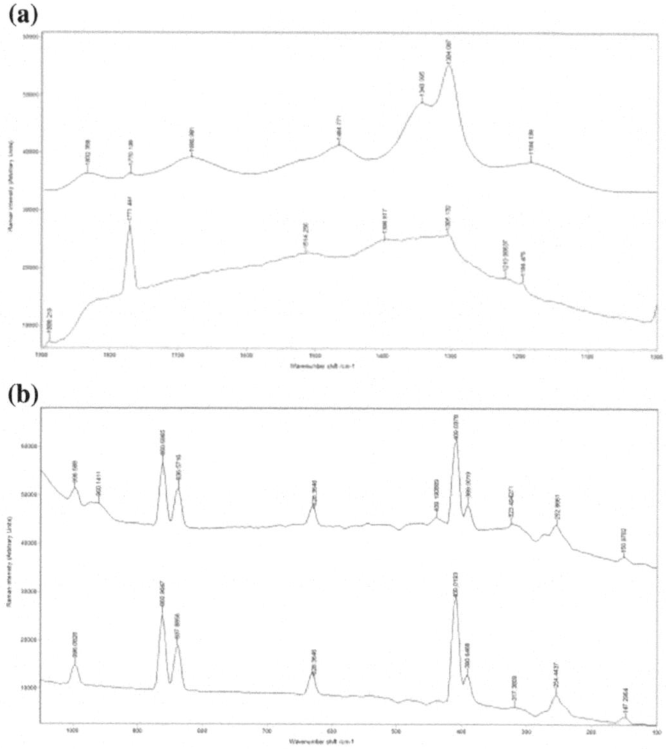

Fig. 6.35 Raman spectra of a Barr, Flight & Barr coffee can (shown in Fig. 6.31). **a** wavenumber region 1850–1050 cm^{-1}; **b** wavenumber region 1050–100 cm^{-1}

the Nantgarw factory, therefore, seems to be upheld by the analytical spectroscopy. Further research on this spill vase following these experiments has revealed that it is probably of Davenport origin (for further details of the research background to this attribution, see Edwards et al., *Raman Spectroscopy in Archaeology and Art History*, 2018).

6.13 Expanded Wavenumber Range, 1050–800 cm^{-1}, Swansea Porcelains

For the three Swansea porcelain body compositions, an expansion of the wavenumber region between 1050 and 800 cm^{-1} in the Raman spectra was undertaken to enhance the observed differences in this critical area, in which the phosphatic components, feldspar and forsterite features appear. The spectra are shown in Fig. 6.37 (Swansea saucer, duck-egg porcelain), 6.38 (Swansea teabowl, glassy porcelain) and

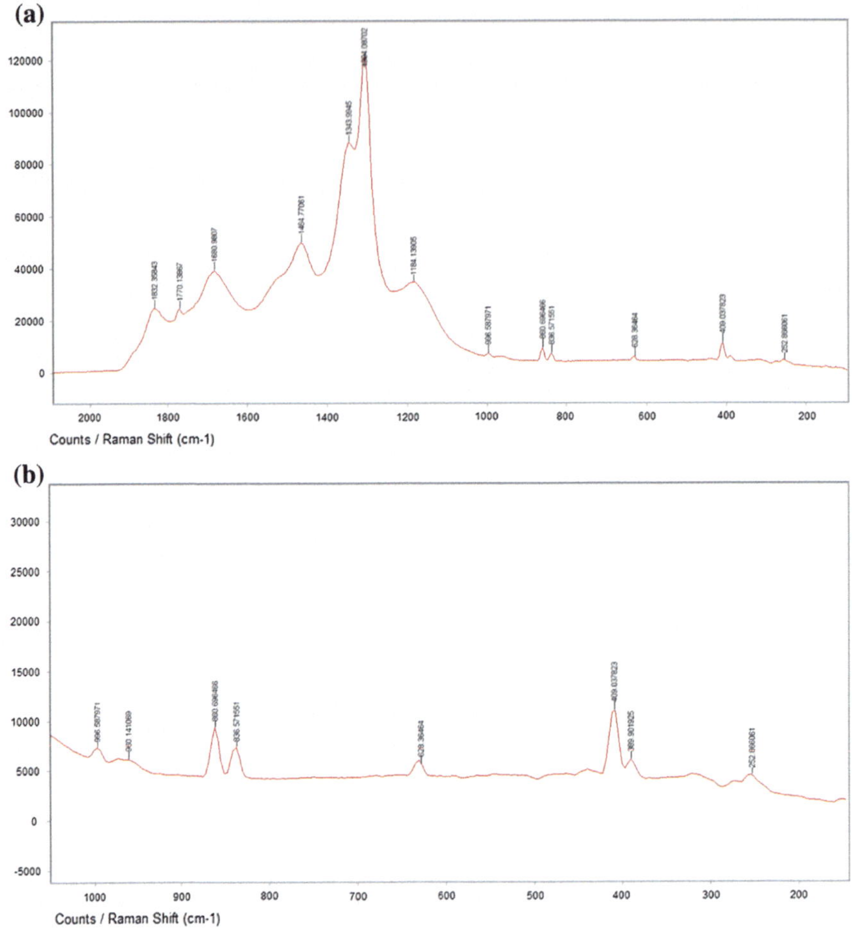

Fig. 6.36 Raman spectra of spill vase, unknown factory, (shown in Fig. 6.32). **a** wavenumber region 1850–1050 cm^{-1}; **b** wavenumber region 1050–100 cm^{-1}

6.39 (Swansea, trident soaprock porcelain) and the bands observed there have been tabulated in Table 6.4.

The spectral differences can be observed and have been highlighted in bold type in Table 6.4, which now provides a methodology and spectroscopic protocol for differentiating between the three types of Swansea porcelain body.

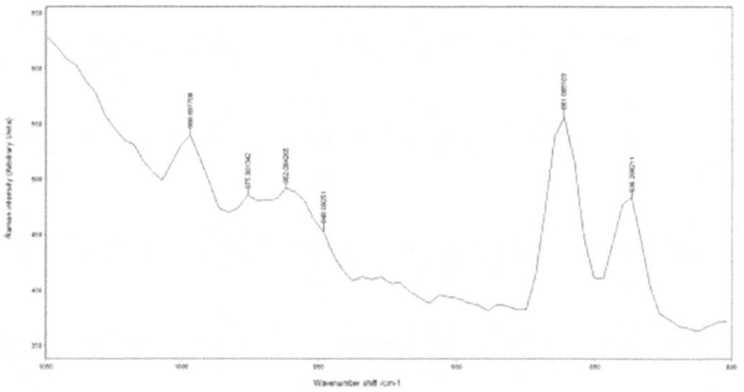

Fig. 6.37 Expanded wavenumber region, 1050–800 cm^{-1}, of the Raman spectrum of the finest Swansea duck-egg porcelain, the deep soup dish shown in Fig. 1.2, marked with SWANSEA in red stencil script

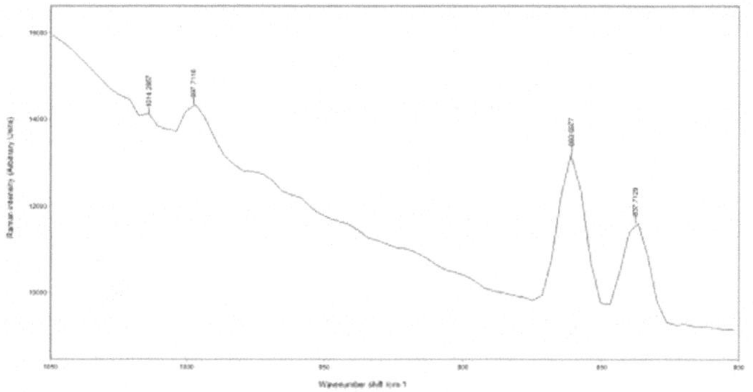

Fig. 6.38 Expanded wavenumber region, 1050–800 cm^{-1}, of the Raman spectrum of the glassy Swansea porcelain tea bowl (Fig. 6.23), unmarked

6.14 Summary

In summary, therefore, the analytical data from 5 Nantgarw shards, 4 pieces of Nantgarw porcelain, 8 pieces of Swansea porcelain and several pieces of porcelain of unknown origin but which are classified as being potentially Nantgarw or Swansea on appearance are reported. This represents the largest collection of perfect and decorated porcelain pieces from Nantgarw and Swansea ever analysed spectroscopically or otherwise and all the analyses have been accomplished successfully and non-destructively for the first time. There is a need to evaluate the analytical return of information against the operational non-destructive protocol: several pieces are of museum quality and are very rare, such as the Swansea violeteer and watering-can, the Nantgarw armorial service crested dinner plate, Nantgarw spill vases, a Swansea chinoiserie spill vase and two potentially unique items in the form of a platter (which appears) to be of Coalport origin but decorated, marked and sold at Swansea, and a

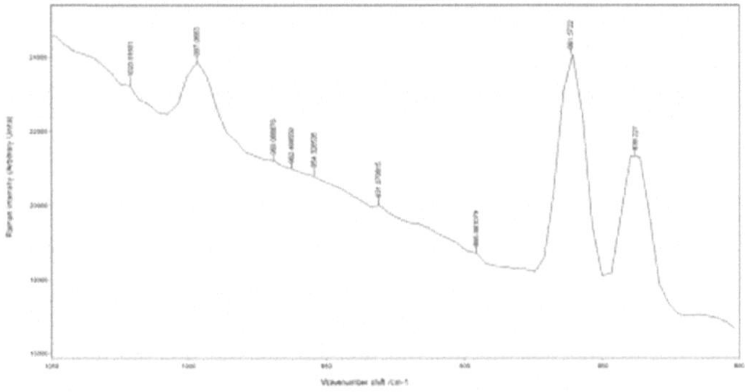

Fig. 6.39 Expanded wavenumber region, 1050–800 cm^{-1}, of the Raman spectrum of the Swansea trident porcelain dessert plate (Fig. 1.6), impressed with SWANSEA and a trident rebus

Table 6.4 Raman wavenumbers for Swansea porcelain specimens: enlarged wavenumber region 1050–800 cm^{-1}

Specimen	Description	Spectral figure	Raman bands/cm^{-1}
9	Duck-egg soup dish	54	1020 vw, 997 m, 976 mw, 962 m, **955 w sh, 948 mw sh**[b], 861 s, 837 ms[a]
8	Glassy teabowl	55	1014 w, 998 m bd, **980 w sh,** 965 w bd, 861 s, 837 ms[a]
14	Trident dessert plate	56	1020 vw, 997 m, 975 vw, 962 vw, **931 w, 895 w**, 861 s, 837 ms[a]

[a]The forsterite doublet, assigned to the symmetric and asymmetric Si–O stretching modes of SiO_4 in Mg_2SiO_4, has the same band relative intensity in all three specimens, 9, 8 and 14, I_{861}: I_{837} of 3:2

[b]These features highlighted in bold type represent the spectroscopic discriminatory reference for the three types of Swansea porcelain studied here

trumpet spill vase which have certain positive attributes for a Swansea or Nantgarw classification of origin, respectively. An important conclusion that can be drawn from Table 6.3 is that Swansea duck-egg porcelain contained glass cullet, which Nantgarw did not, and that Swansea trident porcelain contained soaprock (steatite; Cornish stone: *petuntse*) which neither Nantgarw nor Swansea duck-egg porcelains contained. Both Swansea and Nantgarw contained calcined bone ash, but in varying amounts as a component for different porcelain bodies. Analytically, this offers a significant discriminator between Nantgarw and Swansea porcelains in that the latter should exhibit spectral signals of glass silicates that the former does not: in molecular spectroscopic terms, a new feature should appear for the glassy silicates between 1500 and 1100 cm^{-1} due to Si–O vibrations, and also in the low wavenumber region signatures of Pb–O stretching vibrations near 120 cm^{-1} should appear from the flint glass cullet additive. Elementally, a significantly larger quantitative presence of silica should be noted with the appearance of new signal for lead from the glass cullet.

6.15 Creation of a Raman Spectroscopic Protocol for the Non-destructive Analytical Discrimination Between Swansea and Nantgarw Porcelains, and Porcelains from Some Contemporary Factories

It would be appropriate after consideration of the detailed analytical information made available from the series of experiments carried out here if a working protocol could be established to facilitate the discrimination between Swansea and Nantgarw porcelains, and to afford scientific input to the attribution of specimens which possess several attributes of Welsh porcelains but whose source is uncertain or unknown. This is especially important where porcelains have been faked, or where they are being passed off as genuine examples of Swansea or Nantgarw porcelains which are unmarked or which possess a rare factory mark. From the Raman spectral data, correlated with known compositional recipes for both Nantgarw and Swansea porcelains, it is possible to make the following statements which form the basis for an operational protocol for the identification of unknown or suspect porcelains:

- Swansea porcelain, unlike Nantgarw, contained powdered glass frit and this is manifest in additional bands due to silicates in the higher wavenumber region, which unfortunately are masked by the lanthanide complex electronic spectral excitation features here with 785 nm laser excitation. However, the spectral bandwidths of the modes in Swansea porcelains are slightly larger than those in Nantgarw, which reflect the presence of this component.
- The spectroscopic protocol for the differentiation between Nantgarw and Swansea porcelains therefore depends upon signatures and features in the 1050–100 cm^{-1} region, and these are tabulated in Table 6.2. The key features are the complex bands near 980 and 960 cm^{-1}, which are assignable to wollastonite and whitlockite type species, reflecting differential compositions of bone ash and steatite, a doublet band feature at 830/860 cm^{-1} characteristic of the forsterite/fayalite system, and the bands arising from iron oxides near 400 cm^{-1}. Subtle differences are apparent in the stackplots of the Raman spectra which can be used to discriminate between Swansea an Nantgarw bodies in this wavenumber region.
- The three Swansea bodies are easily differentiated also by reference to the bands in the same region—glassy, duck-egg and trident, especially in the expanded wavenumber plots shown here.
- Additives are not identified in the spectra, so the presence of smalt or borax remains undetected.

References

B.T. Bowie, D.B. Chase, I.R. Lewis, P.R.Griffiths, in *Anomalies and Artifacts in Raman Spectroscopy*, ed. by J.M. Chalmers, P.R. Griffiths in Handbook of Vibrational Spectroscopy (John Wiley and Sons, Chichester, UK, 2002), pp. 2355–2378

E.A. Carter, S.J. Kelloway, N. Kononenk, R. Torrance, Raman spectroscopic studies of obsidian. Anal. Archaeol. 318–344 (2012)

E.A. Carter, M.L. Wood, D. de Waal, H.G.M. Edwards, Porcelain shards from portuguese wrecks: raman spectroscopic analysis of marine archaeological ceramics. Herit. Sci. **5**, 17 (2017). https://doi.org/10.1186/s40494-017-0130-9

P. Colomban, Raman spectrometry: a unique tool to analyze and classify ancient ceramics and glasses. Appl. Phys. A **79**, 167–170 (2004)

P. Colomban, A case study: glasses, glazes and ceramics, in *Raman Spectroscopy in Archaeology and Art History*, ed. by H.G.M. Edwards, J.M. Chalmers (Royal Society of Chemistry Publishing, Cambridge, 2005)

P. Colomban, The destructive/non-destructive identification of enameled pottery, glass artifacts and associated pigments—a brief overview. Arts **2**, 77–110 (2013)

P. Colomban, G. Segon, X. Faurel, Differentiation of antique ceramics from the Raman spectra of their coloured glazes and paintings. J. Raman Spectrosc. **32**, 351–360 (2001)

P. Colomban, V. Milande, H. Lucas, On-site Raman analysis of medici porcelains. J. Raman Spectrosc. **35**, 68–72 (2004)

P. Colomban, F. Treppoz, Identification and differentiation of ancient and modern european porcelains by Raman macro- and micro-spectroscopy. J. Raman Spectrosc. **32**, 93–102 (2001)

A. Cox, A. Cox, *Rockingham Porcelain* (Antique Collectors Club Publishing, Woodbridge, Suffolk, 2005)

P. Craddock, *Scientific Investigation of Copies* (Fakes and Forgeries, Butterworth-Heinemann, Oxford, 2009), pp. 201–210

P. Colomban, Differentiation of antique porcelains, celadons and faiences from the Raman Spectra of their bodies, glazes and paintings. Asian Chem. Lett. **5**, 133 (2001)

L.W. Dillwyn, *Notes on the Experimental Production of Swansea Porcelain Bodies and Glazes*. Made by Lewis Weston Dillwyn with Samuel Walker at the Swansea China Works Between 1815 and 1817. Presented to the Library of the Victoria &Albert Museum, South Kensington, London by John Campbell in 1920. Reproduced in Eccles & Rackham, *Analysed Specimens of English Porcelain*, 1922, see reference

H. Eccles, B. Rackham, *Analysed Specimens of English Porcelain in the V&A Museum Collection* (Victoria & Albert Museum, South Kensington, London, 1922)

H.G.M. Edwards, Swansea and Nantgarw Porcelain Bodies Based on Analytical Evidence: A Case Study, in *Encyclopaedia of Analytical Chemistry*, ed. by R. Meyers, Y. Ozaki (Chichester, UK, J. Wiley and Sons, 2015a)

H.G.M. Edwards, in *Ancient Inks: A Forensic Historical Perspective,* ed. by W.J. Rink, J. Thompson, A.J.T. Jull, J.B. Paces, L. Heamann Encyclopaedia of Scientific Dating Methods (Springer, Heidelberg, Germany, 2015b), pp. 48–52

H.G.M. Edwards, *Swansea and Nantgarw Porcelain: A Scientific Reappraisal* (Springer Publishing, Dordrecht, The Netherlands, 2017a)

H.G.M. Edwards, *Swansea Porcelain: The Duck-Egg Translucent Vision of Lewis Dillwyn* (Penrose Antiques Ltd., Short Guides, Thornton, West Yorkshire, UK, 2017b). ISBN 9780244325787

H.G.M. Edwards, *Derby Porcelain: The Golden Years, 1780–1830* (Penrose Antiques Ltd., Short Guides, Thornton, West Yorkshire, UK, 2017c)

H.G.M. Edwards, in *Nantgarw Porcelain: The Pursuit of Perfection,* ed. by M.D. Denyer, Penrose Antiques Ltd. Short Guides, (Penrose Antiques Ltd., Thornton, West Yorkshire, UK, 2017d). ISBN: 978-0-244-90654-2

H.G.M. Edwards, A.P.H Surtees, R. Telford, in *Dancing on Eggshells: A Holistic Analytical Study of a Ballet Dancer on Regency Porcelain*, ed. by H.G.M. Edwards, P. Vandenabeele. Raman

Spectroscopy in Archaeology and Art History (Royal Society of Chemistry Analytical Chemistry Spectroscopy Monographs Series, Royal Society of Chemistry Publishing Cambridge, UK, vol II, to be published, 2018)
H.G.M. Edwards, M.C.T. Denyer, *William Billingsley The Enigmatic Porcelain Artist, Decorator and Manufacturer* (Penrose Antiques Ltd., Short Guides, Neopubli, Berlin, 2016). ISBN 978-3-7418-6802-3
H.G.M. Edwards, P. Colomban, B. Bowden, Raman spectroscopic analysis of an english soft-paste porcelain plaque-mounted table. J. Raman Spectrosc. **35**, 656–661 (2004)
H.G.M. Edwards and J.M. Chalmers, in *Raman Spectroscopy in Archaeology and Art History*, eds, Royal Society of Chemistry Analytical Chemistry Spectroscopy Monographs Series, (Royal Society of Chemistry Publishing Cambridge, UK, 2005)
N.F. Foster, P.J. Wozniaklewicz, M.C. Price, A.T Kearsley, M.J. Burchell, Identification by Raman spectroscopy of the Mg–Fe content of olivine samples after impact at 6 km s^{-1} onto aluminium foil and aerogel: in the laboratory and in wild-2 cometary samples. Geochimica et Cosmochimica Acta **121**, 241–256 1993
N.D. Gent, *The Patterns and Shapes of the Pinxton China Factory* (Non-Fiction Ceramics, London, 1996)
M. Hillis, The Development of Welsh Porcelain Bodies, in *Welsh Ceramics in Context, Part II*, ed. by J. Gray (Swansea, Royal Institution of South Wales, 2005), pp. 170–192
W.D. John, *Nantgarw Porcelain* (Ceramic Book Co., Newport, 1948)
W.D. John, *William Billingsley* (Ceramic Book Co., Newport, 1968)
A.E. Jones, Sir L. Joseph, Swansea Porcelain: Shapes and Decoration, (D. Brown and Sons, Ltd., Cowbridge, 1988)
D.A. Long, *The Raman Effect: A Unified Treatment of the Theory of Raman Scattering by Molecules* (John Wiley and Sons, Chichester, 2002)
D. Mancini, C. Dupont-Logie, P. Colomban, On-site identification of sceaux porcelain and faience using a portable Raman instrument. Ceram. Int. **42**, 14918–14927 (2016)
E. Morton Nance, in *The Pottery and Porcelain of Swansea and Nantgarw* (B.T. Batsford Ltd., London, 1942)
J.V. Owen, M.L. Morrison, Sagged phosphatic Nantgarw porcelain (ca. 1813–1820): casualty of overfiring or a fertile paste? Geoarchaeology **14**, 313–332 (1999)
J.V. Owen, J.O. Wilstead, R.W. Williams, T.E. Day, A tale of two cities: compositional characteristics of some Nantgarw and Swansea porcelains and their implications for kiln wastage. J. Archaeol. Sci. **25**, 359–375 (1998)
P. Ricciardi, P. Colomban, B. Fabbri, V. Milande, Towards the Establishment of a Raman Database of Early European Porcelain. E-Preserv. Sci. **6**, 22–26 (2006)
J. Sandon, *Starting to Collect Antique Porcelain* (The Antique Collectors Club, Woodbridge, Suffolk, 1997)
E. Smith, G. Dent, *Modern Raman Spectroscopy: A Practical Approach* (John Wiley & Sons, Chichester, 2005)
J. Taylor, *The Complete Practical Potter* (Shelton, Stoke-upon-Trent, 1847)
M.S. Tite, M. Bimson, A Technological Study of English Porcelains. Archaeometry **33**, 3–27 (1991)
E. Widjaja, G.H. Lim, Q. Lim, A.S. Mashadi, M. Garland, Pure Component Raman Spectral Reconstruction from Glazed and Unglazed Yuan, Ming and Qing Shards: a Combined Raman Microscopy and BTEM Study. J. Raman Spectrosc. **42**, 377–382 (2011)

Chapter 7
Reflections on the Holistic Approach to the Analysis of Nantgarw and Swansea Porcelains

Abstract For other ceramics factories the pattern books and order books are still in existence but this is not the case for Swansea and Nantgarw. However, the detailed work books of Lewis Dillwyn have been lodged in a museum archive and they offer a wealth of detail as to his reasoning underlying the changes he made experimentally to his porcelain body at Swansea between August 1815 and 1817. Taken with some letters and documentation, this affords a useful background to the analytical data from Swansea porcelain. Nantgarw is even more sparse by comparison and there is little to inform the analytical data—but for both factories, therefore, the analyses are important for the interpretation of paste compositional change: it seems that Swansea had three main bodies and Nantgarw only one. This is borne out by the analytical evidence that has been obtained thus far.

Keywords Holistic approach · Swansea · Nantgarw · Dillwyn's work books
Porcelain bodies · Analytical correlation

7.1 The Holistic Approach

The historical documentary evidence associated with the establishment, operations and demise of the Nantgarw and Swansea porcelain manufactories has been the subject of several authoritative and comprehensive investigations over the past hundred and twenty years or so and the chemical analysis has run concurrently for almost one hundred years. Research into the former aspect has naturally been focused upon intra-establishment issues and has also recently been extended into inter-factory comparisons: what is clear is that the operational histories of the two Welsh porcelain factories cannot be considered in isolation but requires an appreciation of the general problems facing their competitors resulting from the prevailing issues of the Georgian and Regency periods which affect all the manufactories. Also, the interchange of skilled personnel and ceramics workforce between factories was not

H. G. M. Edwards, *Nantgarw and Swansea Porcelains*,
https://doi.org/10.1007/978-3-319-77631-6_7

an inconsiderable factor in the successful operation and manufacture of porcelain. Regarding this particular aspect, we need look no further than the start-up of the Swansea China Works, which commenced in 1811 with Lewis Dillwyn's vision and employment of two key operatives from Coalport. Henry Morris has commented that the selection of these particular people was not a good one and the operation consequently failed. In contrast, the collaboration between William Billingsley and Samuel Walker at Nantgarw was successful in manufacturing the porcelain, but failed to attract the requisite funding for their continuation: their subsequent engagement with Dillwyn at Swansea in 1812 resulted in the highly successful outcome of the production of Swansea duck-egg porcelain, some of the finest and most translucent china yet made in Europe (de la Beche 1876). The departure of Billingsley and Walker to re-establish Nantgarw in 1817 even bettered their Swansea enterprise, and Dillwyn's Swansea venture then collapsed within two years.

Clearly, therefore, the historical survey of each factory must consider not only the sourcing of materials and the empirical experimental changes in formulations and recipes undertaken by its owners or proprietors, but also needs to examine the influence of internal and external pressures created by the workforce specifically and by society and their clientele, in general.

In contrast, the chemical analysis of ceramics was in its infancy in the late 19th and early 20th Centuries: up until that time, little was known about the chemical reactions which occurred at the elevated temperature regimes operating inside the kilns. Simeon Shaw is generally credited with being the first to attempt to understand the complex processes of porcelain firing in the early 1840s, although in this book a counter proposal has been advanced to include Lewis Dillwyn in this role some twenty years earlier. A key factor in the production of porcelain was the skill of the kiln master, who was responsible not only for the design and construction of the brick kilns themselves but also the placement of the porcelain items for firing. These would be subjected to a ramping of temperature gradients in the kilns for between 24 and 48 h, during which time the kiln master would gauge their exposure to, it is estimated, a few minutes. The major problem facing chemical analysts was the total or partial destruction of their specimens necessitated by the wet chemical processes involved, which demanded relatively large quantities of material for dissolution in strong mineral acids or alkalis, and replicate analyses for the quantitative determination of metals and non-metals. Instrumental analysis was non-existent, and indeed the first ever determination of the composition of a ceramic specimen by instrumental means, namely X-ray diffraction, was accomplished at the British Museum Research Laboratories in 1959, some 63 years after the discovery of X-rays by Wilhelm Roentgen in 1896. The advent of instrumental techniques applied to ceramics analysis is now well-established, but the taking of excised samples or their pretreatment by use of a deposited or sputtered coating is still a pre-requisite for the analytical procedures in most cases. Even so, the removal of micro-sized samples for instrumental analysis is acceptable in only some cases as long as their removal does not constitute a hazard to the specimen: for example, the removal of a small specimen of porcelain body from the underside of a deep dish footrim may be acceptable to the owner in theory but if the mechanical vibrations accompanying the drilling cause a shattering of the object then all is lost! If the porcelain item is rare, unique and valuable then the risk of permanent damage often outweighs the advantage of acquiring analytical confirmation of its origin through its chemical characterisation. Hence, the acquisition of molecular chemical analytical data through non-destructive and non-invasive methods is attractive, even if the information thereby obtained is only partially complete for the total analytical picture. In this account, Raman spectroscopy has been used for the first time to characterise the porcelains of Swansea and Nantgarw using decorated and completed specimens from private collections, and in several cases involving exceptionally beautiful and rare items.

7.2 Survey of Analytical Work on Welsh Porcelains

Analytically, to date, a summary of the specimens of Nantgarw and Swansea porcelain which have been subjected to analytical examination is provided in Table 7.1. The earliest date quoted here for analysis is that of Sir Church (1894) but, as mentioned earlier, no information is provided about the composition of either of the two Nantgarw or Swansea porcelains he cites in his text—these were from the Lady Charlotte Schreiber Collection in the Victoria and Albert Museum, Kensington, London (known as the South Kensington Museum at that time). Eccles in 1914 did analyse a piece of Swansea porcelain it seems, therefore, for the first time in 1914. This featured in the Loan Exhibition at the Glynn Vivian Art Gallery, Swansea, held to commemorate the centenary of the opening of the Swansea China Works in 1814; the piece analysed was from the *Gibbins* service, blue and white transfer printed porcelain described as duck-egg type but bearing the impressed mark for T&J BEVINGTON. An unusual piece, as it must date chronologically from the takeover of Dillwyn's operation by the Bevingtons in 1820 after closure of the Swansea China Works and the analytical figures suggest that perhaps this was an experimental piece from Dillwyn's era and possibly should not be treated as a standard Swansea duck-egg body composition. Eccles and Rackham later (1922) undertook a comprehensive survey of porcelains in the South Kensington Museum Collection, including two examples of Nantgarw porcelain—one being from the Duncombe tea service, which was painted by William Billingsley and offered by him to Edmund Edwards his landlord at Nantagrw in lieu of rent on the site. Three other specimens were Swansea pieces—of which two were duck-egg and one was of a trident formulation. Of these five specimens, all were destroyed in part or completely for the chemical analyses to be performed (Table 7.1).

Tite and Bimson were the first to utilise instrumental techniques to determine the analytical composition of Nantgarw porcelains in 1991. These were applied to three shards in the British Museum Collection obtained from an unspecified excavation carried out in 1933. Owen et al. repeated the analyses for the same Nantgarw shards in the British Museum Archive in 1998 and extended the study to eight Swansea shards personally found on the waste tips at the Hafod site of the Swansea China Works and also to one piece of finished porcelain from pieces of the *Biddulph* Swansea

Table 7.1 Compilation of specimens analysed of Nantgarw and Swansea Porcelain

			Specimens			
			Nantgarw		Swansea	
Analyst	Year	Location	Shards	Finished	Shards	Finished
Church	1894	V&A Museum				
Eccles	1914	Glynn Vivian LE[a]				1
Eccles et al.	1922	V&A Museum		2		3
Tite et al.	1991	British Museum	3			
Owen et al.	1998	BM + RISW[b]	10		8	1
Owen & Morrison	1999	NCWM[c]	10	1		
Edwards	2017	NCW Trust + PC[d]	5	4		1
Total			28	7	8	6

[a]Glynn Vivian Art Gallery, Swansea, Loan Exhibition, Centenary of Swansea China Works, 1914. Specimen of Swansea *Gibbins* service, *ca.* 1820
[b]British Museum archive from 1933 excavations at Nantgarw site; specimen from Swansea *Biddulph* service at Royal Institution of South Wales, Swansea
[c]Nantgarw China Works Museum: including one specimen of a finished porcelain plate
[d]Nantgarw China Works Trust archive, Tyla Gwyn, Nantgarw, shards from 1990s excavation at site; finished specimens from private collection

porcelain service in the Royal Institution of South Wales Collection at Swansea. Although analytical instrumentation was adopted, this was not non-destructive and a damaged, broken piece of the Biddulph service was utilised for their analysis. Finally, we arrive at the present work, the first reported to use non-destructive and non-invasive analytical instrumentation on Welsh porcelains, in which 5 Nantgarw shards, 4 finished pieces of Nantgarw porcelain, and 8 finished pieces of Swansea porcelain were analysed: during this study a spectroscopic protocol was constructed which facilitated the examination of at least four pieces with a suspected Nantgarw or Swansea origin—of which one was definitely confirmed as Nantgarw, one was rejected as being Nantgarw and two were rejected as Swansea. Of the latter, one was identified as being potentially of Coalport origin but decorated by Henry Morris at Swansea. In summary, therefore, it can be stated that the sum total of analytical information obtained from Welsh porcelain over the last 100 years comprises: Nantgarw—11 shards and 6 finished specimens; Swansea—2 shards and 13 finished specimens, a total of 32 individual specimens analysed. Of this the analyses accomplished in the current study represents a significant part, comprising 5 Nantgarw shards and 4 finished porcelain items, and 8 finished items from Swansea. This is over half of the Nantgarw and Swansea china ever studied analytically historically over the last hundred years up to the present time and additionally provides analytical data for three-quarters of the finished porcelain items studied in the same period, viz., 12 from 18 pieces reported analytically, for which all the other six pieces resulted in damage being caused, admittedly in several cases to already broken pieces.

Detailed comparisons of quantitative analytical data have been made here and elsewhere and several conclusions have been formed about the composition of Nantagrw and Swansea porcelain bodies. However, these should be tempered with the number of specimens subjected to analysis: the total number of 32 pieces analysed, of which just over half comprise finished specimens of porcelain (and, of these, only 12 out of 18 examples were actually perfect, undamaged pieces of porcelain, the remainder being damaged pieces) may not seem representative of the output of the Welsh factories over perhaps a total of three years or so operation. This is certainly true when viewed in comparison with the statement of Dr. John that the Nantgarw factory never applied pattern numbers to its products—and this was made after his personal examination of an estimated 5000 pieces of Nantgarw porcelain during a lifetime's study. Nevertheless, the combined analytical data summarised in Table 7.1 can be used as a starting point to make the detailed conclusions in the following section.

7.3 Conclusions of Analytical Studies

- Table 7.2 gives the chemical composition of the components of the porcelain formulations and recipes used by Lewis Dillwyn and Samuel Walker for their series of experimental trials undertaken between mid-1815 and December, 1817, with the purpose of improving the Swansea body. Also, the formulation for the Nantgarw body allegedly divulged to John Taylor by Samuel Walker and which appeared in the *Complete Practical Potter* in 1847 is used as the basis of calculations for the Nantgarw body. Similarly, the three Swansea glazes and three Nantgarw glazes are also considered in this exercise.
- In Table 7.3 the percentage compositions of the Swansea and Nantgarw formulations and recipes are listed: Table 7.4 gives the elemental oxide compositions of Swansea and Nantgarw porcelain bodies and glazes derived from calculations performed using the information provided in Tables 7.2 and 7.3. Table 7.4 is the definitive listing of the relative proportions of the relevant elemental oxides which have been determined analytically and, therefore, facilitates the comparison of the observed chemical analytical data with the recipes actually used by Dillwyn, Walker and Billingsley.
- On the basis of the current work it appears that Nantgarw did use only one porcelain body formulation during the lifetime of its operation; there is documentation that the glaze formulation was changed for Thomas Pardoe's decorating work into a creamier more mellow variety, but significantly, the production of porcelain in the white had then ceased, so the body remained the same.

Table 7.2 % composition of Swansea porcelain bodies and glazes based upon Dillwyn's trial experiments, 1815–1817

Body no.	Sand	Pearl ash	China stone	Soaprock	Bone ash	SS clay	Norden clay	Lime	Borax	Smalt	Arsenic	Lead	Nitre	Total%
1	49.7	8.3	41.4	0.6										100.0
2	25.3	13.6	20.8	2.6					6.9			30.7		99.9
3	37.0	24.7	37.0	1.3										100.0
4			41.4		27.6	27.6	3.4							100.0
5	48.5	2.4	40.0	9.1										100.0
6	46.8	2.3	38.6	12.3										100.0
7			54.2			33.7		12.1						100.0
8	43.7		49.2	1.6		4.9		0.5						99.9
9			50.0			50.0								100.0
10			25.0		33.3	33.3		8.4						100.0
11			33.3		33.3	33.3								100.0
12			30.4		34.8	34.8								100.0
13			28.0		36.0	36.0								100.0
14	43.8	7.5	36.3	12.5										100.1
15	53.3	6.7	40.0											100.0
16	20.7				41.3	27.6	8.6	14.8		1.7				99.9

(continued)

Table 7.2 (continued)

Glaze no.	Sand	Pearl ash	China stone	Soaprock	Bone ash	SS clay	Norden clay	Lime	Borax	Smalt	Arsenic	Lead	Nitre	Total%
1	11.0		36.9			3.7		11.0	7.4		0.5	25.8	3.7	100.0
2	10.6	0.8	37.7			3.8		10.9	7.1			29.1		100.0
3	12.5	1.0	37.4			3.1		10.9	8.3		0.4	26.5		100.1

Nantgarw body and glaze for comparison

Body no.	Sand	Pearl Ash	China Stone	Soaprock	Bone Ash	SS Clay	Norden Clay	Lime	Borax	Smalt	Arsenic	Lead	Nitre	Total%
1	22.6[a]	3.2			41.9	32.3								100.0

(continued)

Table 7.2 (continued)

Glaze no.	Sand	Pearl ash	China stone	Soaprock	Bone ash	SS clay	Norden clay	Lime	Borax	Smalt	Arsenic	Lead	Nitre	Total%
1	19.7		32.6			4.3		12.2	23.6			6.0	1.6	100.0
2[b]	27.7								22.2			50.0		99.9
	18.5								14.8			66.6		99.9

[a]This data has been compiled from the Nantgarw recipe and formulation quoted by Taylor (The Complete Practical Potter, 1847) and stated to have been provided by Samuel Walker; Dr. John (Nantgarw Porcelain, 1948) has drawn attention to the absence of a flint component although it is known that William Billingsley and Samuel Walker took special care in the selection of clear, high quality Norfolk flints for grinding and admixture with their Lynn sand. The %age value for the sand in this Nantgarw body, therefore, should be considered a composite figure for both flints and quartz river sand—both of which are chemically identifiable as pure silica, SiO_2

[b]Glaze number two has been identified with the Pardoe glaze used in the decoration and completion of residual porcelain remaining at the site by Thomas Pardoe between 1820 and 1823 for sale at auction; in contrast, glaze number 1 is the whiter glaze used by Willaim Billingsley during his production at Nantgarw, 1817–1819. The formulation given in Morton Nance (The Pottery and Porcelain of Swansea and Nantgarw, 1942, page 393) for Nantgarw Glaze Number 2 is: sand 5 parts, borax 4 parts and lead oxide either 1 or 2 parts, a much simpler recipe for the glaze than that given by Walker to Taylor (The Complete Practical Potter, 1847). The first row of compositional data given here for NG2 is for the glaze containing 1 part lead oxide and the second row of data refer to the alternative composition containing 2 parts lead oxide

Table 7.3 Elemental oxide compositions/% in porcelain paste and glaze formulation components

Component	Chemical formulation	SiO_2	Al_2O_3	P_2O_5	MgO	CaO	Na_2O	K_2O	PbO	B_2O_3	Co_2O_3
Sand; flint	SiO_2	100									
Bone ash	$10CaOv3P_2O_5 \cdot H_2O$			42.4		55.8					
Lime	CaO					100					
Soaprock	$3MgO \cdot 4SiO_2 \cdot H_2O$	72.6			23.8						
Pearl ash	$K_2O \cdot CO_2$							68.2			
Glass cullet	$xPbO \cdot SiO_2$	[a]							[a]		
Kaolin	$Al_2O_3 \cdot 2SiO_2$	64.4	35.6								
China stone	$SiO_2 \cdot Al_2O_3 \cdot Na_2O \cdot K_2O \cdot MgO \cdot CaO$	73.8	16.3		0.2	1.8	0.4	4.3			
K Feldspar	$K_2O \cdot 6SiO_2 \cdot Al_2O_3$	64.8	18.3					16.9			
Na Feldspar	$Na_2O \cdot 6SiO_2 \cdot Al_2O_3$	68.7	19.4				11.8				
Borax calcined	B_2O_3									100	
Nitre	$K_2O \cdot N_2O_5$							14.7			
Smalt	$CoO \cdot Al_2O_3$		57.7								42.3

[a]Variable composition from PbO: SiO_2 = 10:90 to 60:40 w/w, representing some 21 and 78% RMM of PbO and 79 and 21% RMM SiO_2, respectively

Table 7.4 Quantitative components of elemental oxides/% in Swansea and Nantgarw porcelain body recipes and glazes as determined from formulations provided by Lewis Dillwyn and Samuel Walker

Body Recipe Swansea (LD)	SiO_2	Al_2O_3	P_2O_5	CaO	Na_2O	K_2O	MgO	Others[a]	B_2O_3	As_2O_3	PbO	Total %
No. 1	80.7	6.8		0.8	1.4	7.5	0.1	2.7				100.0
No. 2	42.6	3.4		0.3	0.6	10.2	0.6	5.0	6.9		30.7	100.3
No. 3	64.4	6.0		0.7	1.2	18.4	0.1	9.2				100.0
No. 4	50.6	17.7	11.7	16.1	1.4	1.8	0.1	0.6				100.0
No. 5	84.6	6.5		0.7	1.3	3.4	2.2	1.2				99.9
No. 6	84.2	6.3		0.7	1.3	3.3	3.0	1.2				100.0
No. 7	61.7	20.8		13.1	1.8	2.3	0.1	0.2				100.0
No. 8	84.4	9.7		1.4	1.6	2.1	0.5	0.2				99.9
No. 9	69.1	26.0		0.9	1.7	2.1	0.1	0.2				100.1
No. 10	39.3	16.0	14.1	27.5	0.8	1.1		0.7				100.1
No. 11	46.0	17.3	14.1	19.2	1.1	1.4	0.1	0.7				99.9
No. 12	44.8	17.4	14.8	19.9	1.0	1.3		0.7				99.9

(continued)

Table 7.4 (continued)

Body Recipe Swansea (LD)	SiO_2	Al_2O_3	P_2O_5	CaO	Na_2O	K_2O	MgO	Others[a]	B_2O_3	As_2O_3	PbO	Total %
No. 13	43.9	17.4	15.3	20.6	0.9	1.2		0.7				100.0
No. 14	79.8	8.9		0.7	1.2	6.7		2.5				99.9
No. 15	82.8	6.5		0.7	1.3	6.3	0.1	2.2				99.9
No. 16	44.0	13.9	17.5	23.1				1.4				99.9
Body Recipe Nantgarw (SW)	43.4	11.5	17.8	23.4		2.2		1.8				100.1
Glazes	SiO_2	Al_2O_3	P_2O_5	CaO	Na_2O	K_2O	MgO	Others[a]	B_2O_3	As_2O_3	PbO	Total %
Nantgarw 1	46.6	6.8		12.8	1.1	1.6		1.5	23.6		6.0	100.0
Nantgarw2 (i)	27.2								22.2		50.0	100.0
Nantgarw 2 (ii)	18.5								14.8		60.0	100.0
Swansea 1	40.6	7.3		11.7	1.2	2.1		3.3	7.4	0.5	25.8	99.9
Swansea 2	40.8	7.6		11.6	1.2	2.2	0.1	0.3	7.1		29.1	100.0
Swansea 3	42.1	7.2		11.6	1.2	2.3	0.1	0.3	8.3	0.4	26.5	100.0

[a] refers to minor chemical components determined in analyses

- The Swansea China Works can be associated with just three formulations of distinctive porcelain body composition: in order of production between 1815 and 1819, glassy, duck-egg and trident. All three can be easily differentiated using Raman spectroscopy.
- A fourth type, apparently identified by Owen et al. (1998) from a waster found at the Hafod site, must be treated with some suspicion since it is an outlier in comparison to the others analysed at that time. A possible reason for this could be attributed to a rogue piece of porcelain perhaps "imported" onto the site by Dillwyn and later broken up and discarded—there is no evidence that Lewis Dillwyn adopted the idea of Robert Bloor at Derby of purchasing and crushing wasters in ton quantities from other porcelain manufactories to utilise in his own porcelains but it is known that proprietors of china works did purchase specimens from their competitors for study and possible emulation. Bloor certainly purchased several pieces of Nantgarw from Mortlock's in London, decorated by James Plant in Robins and Randall's atelier; he used these as templates for his celebrated Lord Ongley service at Derby and then he disposed of the Nantagrw exemplars. Several of the latter were acquired by the Museum of Practical Geology at Jermyn Street, London by Sir Henry de la Beche, and it must be a presumption that the remainder were destroyed and deposited perhaps in a waste tip at the Derby China Works. The point being made is that their location at the Derby China Works waste site does not therefore mean forensically that they originated from Derby. Also, it is recorded that for about six years or so after closure of the Swansea China Works. The former Swansea ceramic artists still worked there painting porcelain from outside factories, especially Staffordshire and Coalport. Wasters from their work may therefore have found their way into a waste tip on site and would be true interlopers.
- The spectroscopic analyses can effectively discriminate between Nantgarw and Swansea porcelain bodies, even though the bone ash component in the finest examples of each is very similar. The major distinction between Swansea and Nantgarw porcelains is that Swansea incorporated a powdered flint glass frit into its paste formulation whereas Nantgarw did not. The presence of minor additives in the Swansea paste has not been detected in the experiments described here. Although this detection capability may at first sight seem rather trivial it is would become important for the identification of any items made and decorated at Swansea with a Nantgarw formulation paste—as has been implied by some previous authors. It has also been noted that there seem to be several examples of Nantgarw porcelain plates in existence decorated by Swansea-based artists such as William Pollard: the question then arises as to how this occurred—an obvious explanation, of course, is that Pollard purchased Nantgarw plates at eth final auction sales and then decorated them at Swansea for sale thereafter. It certainly does not mean as some authors have suggested that Dillwyn adopted the Nantgarw paste formulation for his initial experiments and production at Swansea after engaging Billingsley and Walker to join him in Swansea in 1812. A relevant experiment would be to undertake a non-destructive analysis of such an example to verify that it was in fact a Nantgarw body and not a Swansea paste copy using a similar moulding.

- The spectroscopic analyses carried out here have facilitated a protocol for the detection of potential fake porcelains masquerading as Swansea or Nantgarw origins or to confirm attributions to Welsh porcelains for specimens from unknown porcelain manufactories. Similarly, even marked examples can be checked to see if the mark has been applied to emphasise a bogus or incorrect attribution. Both examples have been evaluated here and a spill vase believed to be a possible Nantgarw specimen was confirmed correctly in this, whereas another was shown not to be so. Likewise, a Swansea duck-egg porcelain spill vase attribution was confirmed but a platter marked SWANSEA was shown to be incorrect and relegated to Coalport origin but decorated by Henry Morris at Swansea, probably being signed off by him accordingly. In the same exercise a potential Swansea coffee cup was shown to be of Coalport origin. This demonstrates that the method van be used as a first-pass screening device to examine porcelain attributions and to confirm or otherwise show evidence for origin in an impartial way.
- Alongside the Swansea and Nantgarw porcelain specimens studied in this work several examples of porcelains form other factories have also been examined: of especial note are those of Pinxton and Derby, both of which have very strong associations with William Billingsley, whose name permeates the history and achievements of Welsh porcelain. A future study of this type would facilitate the objective assessment of the influence Billingsley had upon porcelain formulations and experimental trials in the run-up to his foundation of the Nantgarw China Works. For example, Billingsley did not use glass cullet or frit at Nantgarw for his porcelain formulation, yet he did at Derby and at Pinxton. What was the reason behind this decision?
- There is a controversy over the decoration and painter of the Barry-Barry service—generally credited to the Derby China Works—with several possibilities including William Billingsley. A critical factor is the composition of the paste which would dictate whether or not it belongs to the Duesbury or Bloor Derby periods.
- The construction of spectroscopic calibration plots involving key components such as wollastonite and whitlockite found in high temperature fired soft paste porcelain bodies using the appropriate standard minerals in admixture would afford a quantitative methodology for the determination of accurate percentages of the phosphatic and steatitic components in the porcelain bodies. This would directly facilitate the comparison of porcelain bodies which could reveal the results of Lewis Dillwyn's empirical experimentation on paste composition carried out with Samuel Walker during 1816–1817 in his attempts to improve the Swansea body. This would also feed into the progressive replacement of bone ash and its substitution by soapstone to yield Dillwyn's trident body in the last phase of Swansea production during 1817–1819.
- Further Raman spectral studies of several more specimens of Nantgarw and Swansea porcelains were initiated during this project and spectra were recorded of important service items, such as:

1. Nantgarw, *Duke of Cambridge* service, a dinner plate.
2. Swansea, *Burdett-Coutts* service, a dessert plate.
3. Nantgarw, *Lady Seaton* service, a dinner plate.
4. Nantgarw, *Spence-Thomas* service, a coffee cup.
5. Swansea, William Billingsley decorated, a tea plate.
6. Swansea, Japan pattern, a cream jug.
7. Swansea, *Mandarin* pattern, a tea plate.

These spectra have not been reproduced here but comparisons confirm that they match the standard spectra that have been recorded for Nantgarw and Swansea porcelains defined here.

As a final comment, therefore, the preliminary Raman spectroscopic studies reported in this work have opened up a new avenue for the analytical characterisation of Nantgarw and Swansea porcelains and have set a baseline for the role of molecular spectroscopic studies in the forensic evaluation of unknown porcelains which potentially can be attributed to these Welsh factories.

It was announced in 2017 that the Nantgarw China Works Trust had received financial support from the Welsh Office for the re-creation of a novel porcelain based on the Nantgarw formulation: clearly, it will be appropriate for Raman spectroscopic studies to be performed upon this ceramic body to verify that it is compatible chemically and structurally with the true Nantgarw formulation created by William Billingsley, Samuel Walker and William Weston Young in 1817.

7.4 Detailed Comparisons Between the Recipe Formulations and Elemental Oxide Analytical Data for Swansea and Nantgarw Porcelains

This section concerns the data shown in Tables 7.2, 7.3 and 7.4: these have been derived from the published recipes and formulations described above and the analytical determinations of the minerals and materials used in the porcelain pastes and frits. A problem hitherto which has been identified earlier is that the elemental oxides determined analytically from fired porcelain specimens are often cumulative results from several source materials: for example, SiO_2 is silica and has a presence in china stone, feldspar, sand, flint, soaprock, kaolin and smalt. Likewise, Al_2O_3 is to be found in china stone, kaolin and feldspar and potassium oxide K_2O in feldspar, pearl ash, nitre and china stone. It is really almost impossible, therefore, to make objective and quantitative interpretations about composition of the original porcelain mixture before firing from determinations of these elemental oxides: only in a few cases does the determination of an elemental oxide actually relate exclusively and unambiguously to its original and unique component—examples of this are found for P_2O_5, B_2O_3 and PbO, which directly inform the analyst about the original proportions of bone ash, calcined borax and flint glass cullet actually used in the recipe. Another example, which is less important for compositional discussions, is sulfate, which arises uniquely from a gypsum impurity occasionally found in calcite and limestone and can find its way into the fired porcelain body through contamination of lime derived from calcite or dolomitic limestone.

In each Table, the sum total of the contributing percentages of each component and elemental oxide appear in the final column: the individual percentages have been rounded up in the calculations to the nearest 0.1%, and the sum totals which are usually n the range 99.9–100.1% therefore give a reasonable assessment of the overall accuracy of the method. Likewise, the absence of a percentage for minor components in the Tables can be related to their presence at a level less than 0.05% and anything in excess of this concentration is rounded up to 0.1%. An assumption made in the construction of the breakdown of the components of the original minerals and materials for porcelain manufacture is that these are assumed to be pure and that the minerals themselves have a formulation matching that of their pure chemical formulae. It is appreciated that Dillwyn and Billingsley recognised the vital importance of using pure materials of the highest quality for their porcelain manufacture—resourcing their sand, flints and kaolin from carefully selected localities which invariably were not local to their operations and kiln sites. Nevertheless, the presence of even small impurities will distort the overall picture within the percentage levels cited here. One feature in Table 7.4 is the column marked "Others", which is designed to include minor species such as Fe_2O_3, Co_2O_3, TiO_2, H_2O, NO NO_2 and CO_2 which occur in raw materials at relatively low levels or which are designated components in materials such as nitre, bone ash, soaprock, borax and pearl ash. The determined percentages of metal oxides such as TiO_2 and Fe_2O_3 in china stone, for example, are quoted as 0.15 and 0.2%, which are comparable in percentage terms with magnesium oxide MgO at 0.15%. However, magnesium oxide is cited as a component in the Tables because of its multiple occurrences in other raw materials at higher concentrations, for example, soaprock where MgO occurs to the extent of 23.8%. The volatile components in these raw materials will be removed in pre-preparative production processing such as the preliminary calcination of powdered bone and borax whereas higher temperatures are required to effect decomposition in pearl ash (resulting in the loss of CO_2) and soaprock (resulting in the loss of occluded H_2O). An idea of the importance of including a designated component percentage for these minor components and volatiles is seen in Table 7.4, where percentages of between 2 and 3% are commonly noted. For starting materials of potentially rather indefinite formulation then average literature values from a series of analytical determinations are taken: this is particularly indispensable for bone ash, which is typically cited as 55.8% CaO, 42.4% P_2O_5 and 1.8% H_2O and flint glass cullet which has a variable composition ranging from about 6–60% lead oxide in silica. It is preferable to use these analytical determinations rather than calculations based on strict chemical formulae such as $Ca_5P_3O_{13}H$ (or alternatively $10CaO \cdot 3P_2O_5 \cdot OH$) and $PbSiO_4$ (or alternatively $2PbO \cdot SiO_2$), respectively. In the latter case with the adoption of relative molecular masses of 223.2 for PbO and 92.1 for SiO_2 a lead oxide:silica glass of composition 10% lead oxide and 90% silica will actually have a molar ratio of 1:4 in which the lead component has a 20% ratio: in a 50:50 lead oxide:silica glass,the lead oxide has a 71% molar ratio.

References

Sir A.H. Church, *English Porcelain: A Handbook to the China Made in England During the 18th Century as Ilustrated by Specimens Chiefly in the National Collection* (A South Kensington Museum Handbook, Chapman & Hall Ltd., London, 1885 and 1894)

Sir H.T. de la Beche, T. Reeks, F.W. Rudler, *Catalogue of Specimens in the Museum of Practical Geology, Jermyn Street, London, Illustrative of the Composition and Manufacture of British Pottery and Porcelain from the Occupation of Britain by the Romans to the Present Time*. 3rd ed. by G. Eyre, W. Spottiswoode (For the HMSO, London, 1876)

L.W. Dillwyn, *Notes on the Experimental Production of Swansea Porcelain Bodies and Glazes*. Made by Lewis Weston Dillwyn with Samuel Walker at the Swansea China Works Between 1815 and 1817. Presented to the Library of the Victoria &Albert Museum, South Kensington, London by John Campbell in 1920. Reproduced in Eccles & Rackham, *Analysed Specimens of English Porcelain*, 1922, see reference

H. Eccles, B. Rackham, *Analysed Specimens of English Porcelain in the V&A Museum Collection* (Victoria & Albert Museum, South Kensington, London, 1922)

W.D. John, *Nantgarw Porcelain* (Ceramic Book Co., Newport, 1948)

J.V. Owen, J.O. Wilstead, R.W. Williams, T.E. Day, A tale of two cities: compositional characteristics of some Nantgarw and Swansea Porcelains and their implications for Kiln Wastage. J. Archaeol. Sci. **25**, 359–375 (1998)

M.S. Tite, M. Bimson, A technological study of english porcelains. Archaeometry **33**, 3–27 (1991)

J. Taylor, *The complete practical potter* (Shelton, Stoke-upon-Trent, 1847)

Chapter 8
Critical Evaluation of Dillwyn's Recipes and Formulations

Abstract The work books of Lewis Dillwyn which reveal the series of empirical changes made to the formulations of his Swansea porcelain bodies between August 1815 and 1817 aided by his kiln manager, Samuel Walker, give details of his carefully altered compositions which were carried out on trial systems. Each one has a rather terse statement in which his achievement of a resulting finer porcelain body was pleasurably expressed. It is clear that Swansea never tried to emulate Billingsley and Walker's Nantgarw formulation despite several historical assertions to the contrary. It is also now possible from these workbooks to match the starting compositions with analytical data from the fired porcelain bodies and shards.

Keywords Dillwyn · Work books · Trial compositions · Nantgarw · Formulation recipes · Fired porcelain data

8.1 Critical Analysis of Dillwyn's Formulations for Swansea Porcelain

The recipes and formulations that were undertaken by Lewis Dillwyn are reproduced with comments and transcription of his shorthand codes for the various components used are given in Appendix B. From calculations carried out by the author on this information an estimate has been made of the %age component composition of each trial formulation and these data have been given in Table 7.2, along with the available analogous data derived from the Nantgarw formulation presented by Taylor (*The Complete Practical Potter*, 1847) which he claimed to have originated from Samuel Walker before he departed for the USA to set up his manufactory in Temperance Hill, New Jersey, in the second quarter of the 19th Century.

A close examination of Lewis Dillwyn's work book notes reveals some interesting data and interpretations about experiments on Swansea porcelain variants that took

H. G. M. Edwards, *Nantgarw and Swansea Porcelains*,
https://doi.org/10.1007/978-3-319-77631-6_8

place between 1815 and December, 1817, carried out by Lewis Dillwyn and Samuel Walker:

- The timing of these experiments is very revealing: all were undertaken in the period between 1815 and 1817 when Dillwyn's duck-egg porcelain was created and at the height of its appreciation by the clientele. This is in contrast with some perception in the literature that these experiments were a last resort attempt by Dillwyn to improve the robustness of his finest Swansea duck-egg porcelain, resulting in the trident body that eventually proved to be the downfall of the factory economically. A successful result in the production of a more robust china but sacrificing the translucency, texture and beauty of the original that was commercially disappointing in its unacceptability by society. The recipes describe chronologically the discovery of the duck-egg body in 1815 and eventually the trident variant—however, it is interesting that the first record of the trident body production appears early in the work book in the autumn of 1815, labelled *Body Number 3*, and this curiously must have occurred not long after the first duck-egg body, labelled *Body Number 2*, when presumably the need to develop a robust body was not obviously apparent! A second and third attempt at a trident body is listed later, probably in early 1817, cited as *Body Number 12* and *Body Number 13*, and this interpretation would fit in better with the established credo that the trident body was needed desperately to try to turn around the Swansea China Works ebbing fortunes at that time. Despite this, even in December 1817, with *Body Number 14*, it is evident that Dillwyn was still trying to persevere with a fine translucent china that reverted to his successful duck-egg type formulations, and perhaps he was trying even at that stage to create a more robust china which had the duck-egg translucency for which he was esteemed?
- The listing of component percentages given in Table 7.2 are based upon the recipes and formulations written by Lewis Dillwyn in his work book between 1815 and 1817: fourteen different component items gave been identified from these notes, from the major constituents such as sand, pearl ash, china stone, bone ash, china clay (St. Stephen's Cornwall clay and Norden ball clay) and soaprock, to the minor additives such as arsenic, smalt, borax and nitre. However, most of the Swansea body formulations contained only 4 or 5 components although the three glazes did contain 8 or 9 of those listed.
- Despite the extant contemporary documentation which describes the efforts of china manufactory proprietors to source the finest flints from Norfolk, East Anglia, with but little contamination from iron compounds in particular, in only one recipe is flint mentioned by Dillwyn, namely Body Number 16, dating from his last experimental record in December 1817. William Billingsley also made special reference to the necessity for the fine grinding of flints and components for his frits in his preparation of the Nantgarw paste. However, sand is used in 9 of the 16 body recipes and all 3 of the glaze recipes—and this would presumably have been the finest quality fine quartz sand from the Lynn gravel beds in Norfolk as specified in Walker's Nantgarw revelation to Taylor (The Complete Practical Potter, 1847). Actually, chemically the major constituent of both fine river sand

and flints is identical, namely silica SiO_2, and upon fusion in the ceramic kilns this would provide a suitable matrix of orthosilicates for binding with the other additives and components in the paste mixture. In conclusion, therefore, we can allege that flints were commonly used as components of the porcelain pastes as implied in the associated contemporary documentation but we are not able to define their precise quantitative composition as part of the "silica" percentage. Similarly, the component percentages quoted in Table 7.2 for Nantgarw porcelain show just four constituents, with a zero entry for flints.

- In contrast, Dillwyn does differentiate between his component clays sourced from the St Stephen's and Norden mines near St Austell, Cornwall. He also recognises the chemical difference between china stone and soaprock as has been already described earlier: Eccles and Rackham (Analysed Specimens of English Porcelain, 1922) fail to achieve this distinction clearly in their preamble and actually comment that china stone is analogous to the Chinese petuntse, which it is not as the source mines for each are located some 8000 miles apart.
- Dillwyn describes two experiments which can be best described as outliers, combined in his Body Number 9 notes, wherein only two components are utilised in the body paste, namely china stone and St. Stephen's clay. This manufacture resulted in a hard paste porcelain composite closely resembling Dresden and French chin and would have been very difficult to ascribe or attribute to a Swansea product if examples were found today.
- A very curious entry in Dillwyn's notebook refers to Body Number 3 as a "Variation on the Nantgarw body"; it is suspected that this is seen as the evidential source of much conjecture about the manufacture of the Nantgarw analogue porcelain body at Swansea, especially in the early days of employment there of Billingsley and Walker. When one refers to Table 7.2 and the respective entries for the Swansea Body Number 3 and for Nantgarw as revealed by Walker and re-created by Professor Mellor in the 1880s this just does not ring true at all. For example, the Nantgarw body comprises just four components, namely sand, pearl ash, bone ash and St. Stephen's china clay, whereas the Swansea Body Number 3 contains sand, pearl ash, china stone and soaprock—these are very different mixtures indeed and cannot be remotely be considered similar. In fact, the Swansea paste mixture would be a better conceived forerunner for Dillwyn's trident body rather than the soft paste Nantgarw formulation. It should be noted too that the Swansea porcelain Body Number 3 does not contain any bone ash or china clay at all—and it has long been maintained that the high bone ash content of 41% in Nantgarw was an integral part of its formulation! Examination of the data in Table 7.2 reveals that none of the 16 Swansea body compositions noted in Dillwyn's diaries even approaches a Nantgarw formulation: even those Swansea bodies with high bone ash and china clay %ages such as Bodies Number 7, 10, 11, 12 and 13, all have a china stone or soaprock content but do not exhibit any sand, flint or pearl ash component possessed by their Nantgarw exemplar.
- The Swansea glazes, Numbers 1, 2 and 3, in contrast, are apparently similar to that of Nantgarw glaze Number 1, but quantitatively differ enormously: the Swansea lead content is generally significantly larger than that of the Nantgarw glaze but

the borax and sand content are very much smaller, although the china stone, lime and china clay contents are similar. Finally, the Nantgarw glaze Number 1 does not contain any pearl ash, unlike its Swansea analogues. Nantgarw glaze Number 2, which is accredited with the formulation used by Pardoe and Young in their decoration of the residual Nantgarw stock at the factory prior to its disposal through auction in 1820–1823, has only three components, namely sand, china stone and a very high proportion of lime. Often thought to be of a rather yellow cast, the Pardoe glaze on Nantgarw china does not contain any lead oxide, borax, china clay or nitre unlike its Billingsley analogue, and must therefore be considered to be rather more infusible, despite its higher alkali calcareous flux content which does not replace the high borax %age component in Billingsley's glaze.

- Eccles & Rackham in their classic text on porcelain analysis (Analysed Specimens of English Porcelain, 1922) state an opinion that Dillwyn's Body Number Three and Body Number 12 are a first and second attempt, respectively, at the creation of his trident body: this is rather difficult to appreciate in an evaluation of the data in Table 7.2 as their compositions are so diverse. Body Number 3 contains a high proportion of sand, china stone and pearl ash with a much smaller quantity of soaprock whereas Body Number 12 contains no sand, soaprock and pearl ash and has only half the quantity of china stone with equally significant amounts of bone ash and St. Stephen's clay! Clearly, these comprise very different compositions and cannot be considered both as derivatives of a trident body. The author would suggest that on the basis of the data presented in Table 7.2 derived from Dillwyn's experiments that Body Number 14 would better fit a trident paste—comprising significant amounts of sand, china stone and soaprock with a smaller quantity of pearl ash—in fact, this body is a clear derivative of the earlier Body Number 5 and Body Number 6 which both have similar proportions of sand, china stone and soaprock to Body Number 14. By late 1817, which is the last date in Dillwyn's notebook, the pressing need for a robust Swansea body was starting to be obvious to Dillwyn and this could well explain his reversion to the soaprock/china stone combination of his earlier experiments and the creation of Body Number 14 as a result.
- Eccles and Rackham (Analysed Specimens of English Porcelain, 1922, page 15) also make reference to Dillwyn's Swansea Body Number 8 as being a very close approximation to the Nantgarw body. Comparison of the data for the Nantgarw body and Body Number 8 in Table 7.2 shows that this suggestion is not tenable: the Nantgarw body contains sand, bone ash and St. Stephen's clay in major proportions with a minor component of pearl ash whereas Swansea Body Number 8 has a major component of sand (almost 3.5 times that of its Nantgarw analogue) and only a minor presence of china stone, soaprock, St. Stephen's clay and lime—the pearl ash and bone ash are not present at all in the Swansea body and the Nantgarw body does not contain china stone, soaprock nor lime!! We can but conclude that these two bodies are chemically different. In fact, not one Swansea body of the 16 trialled by Dillwyn between 1815 and 1817 can be considered to be remotely similar to a Nantgarw analogue.

- An interesting range of experimental compositions can be noted for Swansea Body Numbers 10,11,12 and 13 which can be selected as a subgroup of only three components, namely china stone, bone ash and St. Stephen's clay wherein Dillwyn has varied the proportions slightly for the bone ash and clay to the extent of 20% (range 30.8–43.4%, average 35.9%) for each component, whilst maintaining the amounts of these as a constant and changing the amount of the china stone component over a larger range from 16.4 to 30.8%, average 27.1%. For Body Number 10, Dillwyn maintained the same proportions of china stone, soaprock and St. Stephen's clay at 30.8% each, whilst adding a small amount of lime, presumably to aid better alkaline fusion. Then for Body Number 11, he omitted to add the lime and built up the proportions equally of the china stone, soaprock and the St. Stephen's clay. Then he decided for Body Number 12 to halve the quantity of china stone and increase the proportions of the soaprock and clay equally by 33%. Finally, for Body Number 13 Dillwyn reduced the china stone by only 15% and then increase the soaprock and clay proportions equally by only 10%. These experiments were carried out in eth period between the Autumn, 1816 and March, 1817: Dillwyn expressed his satisfaction with the fine quality of the porcelain in Body Number 10: "... A beautiful china which stands well but is rather too soft for the hard glaze". In Body Number 11 he was pleased with the result of the omission of lime and the consequent proportional increase in the china stone, soaprock and St. Stephen's clay components, commenting thus "... a beautiful body and in all respects answers". Further tuning of the relative proportions of the three components of china stone, soaprock and St Stephen's clay Dillwyn found to be a further improvement until finally he achieved his desired porcelain hard body with Body Number 13, his prototype trident body .... "... makes the body harder but large pieces are more apt to fly" and was especially pleased with the quality of the glaze on this body with the application of Glaze Number 2. Thereafter, Dillwyn turned his attention from the trident body and attempted to find a more robust duck-egg variation which he seems to have achieved with Body Number 15 and Body Number 16 which he described as a beautiful and good body, Number 16 being little improved upon Number 15.

8.2 Correlation Between Dillwyn's Experimental Data and Analyses of Swansea Porcelain

As stated earlier, the major problems associated with the correlation of recipe formulations and use of original components and the analytical data accumulated from the interrogation of finished, fired porcelain pieces are the chemical changes which have occurred in the high temperature kilns. Even so, as we have seen from Lewis Dillwyn's notebooks, the usage of original starting materials is often subject to potential mis-interpretations: classic examples include the substitution of flints for fine sand and the addition or otherwise of lead glass to the porcelain mixture. It is fortunate that analysts have two unequivocal elemental markers and parameters in phosphorus and lead—the former being an indicator exclusively of the use of bone ash and the latter of flint glass in the porcelain frits prepared for mixing as pastes. A minor elemental indicator is sulfur, which registers the presence in the paste of gypsum in its hydrated or anhydrous forms mineralogically, namely gypsum, anhydrite or bassanite, in which the $CaSO_4$ has associated with it (either 2, 0 or $1H_2O$) molecules.

An example of this difficulty of correlation between the analytical data and the original components can be cited with silica, SiO_2, which occurs in several components commonly utilised in the preparation of porcelain body pastes: river sand, flints, glass and smalt are all silicates which are based on quartz structures. Further silicates are to be found in clays and china stone, as well as in talcs and soaprock. All these silica and silicate components convert to high temperature forms in the kilns and even react to form mixed compounds with phosphates in the presence of bone ash. Alkaline components which are added to increase the fusibility or flux of the incipient porcelain in the kiln, such as pearl ash, borax and lime, can react with silicate structures at elevated temperatures. Nevertheless, much detailed analytical interpretation has been forthcoming form porcelain analyses which until now have had a major disadvantage in generally being destructive of part or whole of the artefact concerned. In this text, a way forward has been advanced for the adoption of non-destructive analytical techniques which can interrogate a finished and decorated specimen through the glaze without any form of mechanical or chemical pretreatment being undertaken.

The correlation between the actual elemental oxide components of the initial porcelain paste formulation composition and the relevant analytical data determined from the fired ceramic materials it is necessary to evaluate the contributions made from specific oxides to the original materials: normally, this is accomplished scientifically through the calculation of the relative molar masses of each oxide and that of the whole compound material, expressing these data as molar percentages. This is straightforward for simple and well-structured chemical formulations such as quartz and sand, which comprise just SiO_2 as silica with insignificant impurities present such as iron (III) oxide, so sand can be represented as 100% silica. Other minerals, synthetic materials or natural rocks used in the porcelain paste mixture, however, are not so easily transformed into the ratios of their chemical formulaic components, such as china stone, smalt and glass cullet. Table 7.3 provides a quantitative formulaic listing of the common ingredients in porcelain composition mixtures in which the %ages of the components have been calculated on the basis of the relative molar masses of the constituents: to achieve this the relative molar masses of the key oxide components have been determined with the following results—SiO_2 92.1 Da, Al_2O_3 102 Da, P_2O_5 122 Da, MgO 40.3 Da, CaO 56.1 Da, Na_2O 62 Da, K_2O 94.2 Da, PbO 223.2 Da, B_2O_3 69.67 Da, and Co_2O_3 165.8 Da. Using these values, the relative percentage concentrations of each of these oxides in the starting materials has been determined and the data listed in Table 7.3.

8.3 Gallery of the Key Players in the Creation of Welsh Porcelains

Aside from the scientific and historical themes covered in this text regarding the two Welsh porcelain factories which inform the analytical determinations and elemental compositions of the porcelain bodies, it is appropriate to consider the key personnel associated with Nantgarw and Swansea porcelain manufacture to enable readers to better assimilate the personalities of the people behind the names that have been mentioned throughout this work, namely: William Billingsley, Lewis Weston Dillwyn, William Weston Young, Samuel Walker and Thomas Pardoe.

William Billingsley: 1758–1828. The only physical description we have of William Billingsley is that mentioned in Turner's book (The *Ceramics of Swansea and Nantgarw*, 1897), where an interview with Richard Millward, a former worker at the Nantgarw China Works, revealed that Billingsley was "a *thin man of middle height, fair, with grey hair, had no beard: he was a pleasant speaking man, but very hot tempered*". Clearly, he did not suffer fools gladly. A line drawing originally ascribed to William Corden but now reassigned to "The Rose Painter", aka William Billingsley, aged about 37, just before he left the Derby China Works in 1795 is shown in John Twitchett's book (Derby *Porcelain*, 1980, plate 252, page 202): it is easy to see that Millward's description is compatible with the essence of this painting. Billinglsey died in Coalport in 1828 and it now seems to be generally accepted that neither he nor Samuel Walker ever revealed publicly the Nantgarw formulation during Billinglsey's lifetime. However, John Taylor (in the *Complete Practical Potter*, 1847) does suggest that Walker did instruct him in the details of the Nantgarw body formulation probably before departing for the USA to set up the Temperance Hill Pottery in New Jersey.

Lewis Weston Dillwyn: 1778–1855. After selling his interests in the Swansea China Manufactory to Timothy and John Bevington, who oversaw the disposal of all remaining Swansea porcelain at the auction sales between 1823 and 1826, thereafter he seemed to lose interest in porcelain manufacture.

William Weston Young: 1776–1847. Young was a Quaker and refused to have his portrait recorded for posterity. After the departure of Billingsley and Walker from Nantgarw to join John Rose in Coalport in 1820, Young engaged Thomas Pardoe then based in Bristol to join him in Nantgarw to decorate porcelain remaining in the white for sale at auctions—contrary to some opinion, no porcelain was made at Nantgarw in this period, and after only a short while, Thomas Pardoe died in July 1823, and he is buried there in Eglwyssilan Church. It can be inferred that possibly Walker did approach Young sometime after 1828 to see if the Nantgarw factory fortunes and production could be re-instated, but by that time the impetus was lost.

Thomas Pardoe: 1770–1823. A self-portrait ca. 1810–1820 exists of Thomas Pardoe in The National Museum of Wales, Cardiff, where it was presented by his descendants in 1959. It shows a middle-aged man holding a china vase—although an earlier version shows him to be, like Billingsley, an ascetic. The Pardoe family home situated alongside the Glamorgan Canal at Nantgarw, was still occupied by eth family after Thomas' death in 1823 and William Pardoe did commence the manufacture of clay pipes at the Nantgarw China Works site in 1835 until commercial enterprise and the marketing of ready-rolled paper cigarettes by the American Tobacco company in the 1880s sounded the death knell for clay pipe manufacture. However, the "continuance" of production of earthen ware clay pipes at Nantgarw after 1823 has given rise to several misinterpretations in earlier literature concerning the re-invigoration of porcelain manufacture there, which never happened.

Samuel Walker: 1790–1880. No picture exists of this important role player in Nantgarw and Swansea porcelain manufacture. He was Billingsley's son-in-law and was married to Sarah Billingsley—faithfully assisting materially in the porcelain travails of the Billingsley family from Brampton, through Worcester, on to Nantgarw, then Swansea and back to Nantgarw, finally ending up with William Billingsley in Coalport, where he is credited with the invention of a superior claret ground colour for adoption in their workshops. His role at Coalport is not clear: Walker was a kiln designer and manager and not a chemist as claimed by others, although he did ably assist Lewis Dillwyn in the achievement of his esteemed duck-egg porcelain body at Swansea in 1817. Clearly, he also did experiment at Coalport in the creation of novel ground enamel colours, probably assisting Billingsley in his role as artistic adviser to John Rose's atelier there. In the 1840s he emigrated to new Jersey and was instrumental in setting up a pottery manufactory at Temperance Hill, form which several reasonably high pieces have been recently identified—although it seems that porcelain manufacture was not attempted.

The fact that so few pictures exist of early figures and personalities from 18th and 19th Century porcelain manufacture can be attributed to a time well before photographic portraiture had gained a foothold, which occurred only after the ground-breaking work of William Fox Talbot on the calotype photographic process patented in February, 1841.

References

L.W. Dillwyn, *Notes on the Experimental Production of Swansea Porcelain Bodies and Glazes.* Made by Lewis Weston Dillwyn with Samuel Walker at the Swansea China Works Between 1815 and 1817. Presented to the Library of the Victoria &Albert Museum, South Kensington, London by John Campbell in 1920. Reproduced in Eccles & Rackham, *Analysed Specimens of English Porcelain*, 1922, see reference

H. Eccles, B. Rackham, *Analysed Specimens of English Porcelain in the V&A Museum* (Victoria & Albert Museum, South Kensington, London, 1922)

J. Taylor, *The Complete Practical Potter* (Shelton, Stoke-upon-Trent, 1847)

W. Turner, *The Ceramics of Swansea and Nantgarw* (Bemrose & Sons, Old Bailey, London, 1897)

Chapter 9
Perspective of Analytical Science in Ceramics Research

Abstract The adoption of analytical science to characterise the output of a porcelain manufactory has hitherto been directed at shards or wasters from factory sites and badly damaged finished pieces because of the destructive sampling regimes required: however, the advent of noninvasive and nondestructive analytical techniques have brought a new dimension to this science because for the first time it has become possible to analyse finished and perfect pieces of porcelain nondestructively without the need for sampling, chemical or mechanical pretreatment. Future studies will involve the extension of a current analytical database to include more factories and also to probe different periods of their manufacture to investigate novel body and perhaps glaze compositions. Portable analytical instrumentation will facilitate the access to museums and private collections where rare and documentary pieces are held.

Keywords Shards · Non-destructive sampling · Body compositions · Portable analytical instrumentation · Rare porcelain

The title of this book includes the phrase "an analytical perspective" and much emphasis has been laid upon the role of analysis in holistic research, in which several examples of its successful application have been cited; this approach has adopted type of forensic art investigation, which has necessarily examined historical documentation and, more importantly, the statements that have been made by authors often a century or so ago which may now be regarded as imprecise or even misleading in the light of modern research outcomes. At first sight the role of chemical analysis in determining the composition of the bodies or glazes of porcelains may seem irrefutable and a defining factor for the potential attribution of a factory of origin to a piece of ceramic art work, and the correlation of modern analytical data with historically documented recipes and formulations may seem to be straightforward. However, although the interpretation of the chemical analytical data and its comparison with the raw starting materials ostensibly used in the manufactories is in itself

H. G. M. Edwards, *Nantgarw and Swansea Porcelains*,
https://doi.org/10.1007/978-3-319-77631-6_9

not without its inherent problems because of the high temperature transformations that have occurred in the kiln firing processes, the major stumbling block appears to reside in the correlation between the porcelain body compositions and potential conflict with the expert opinions that have already been expressed regarding factory origins and belief based upon stylistic, textural, decorative and spatial factors.

9.1 The Curious Case of Bow

A classic example of this mismatch is demonstrated by the extensive and comprehensive analytical work using scanning electron microscopy and energy dispersive X-ray spectrometry carried out in the last decade or more by Ramsay and Ramsay (2007a, b), who have presented detailed qualitative and quantitative analytical data for some early Bow porcelains, from which they have concluded that some of the earliest attempts to make porcelain at Bow have gone unrecognised hitherto. As a result, they have argued that the standing of the Bow factory in the history of English porcelain manufacture should be re-assessed to stress its importance and that these early trials in porcelain manufacture were conceptually leading in Europe, outweighing the experiments in porcelain production being carried out at Meissen in Saxony. The analyses of Ramsay & Ramsay clearly demonstrate that the Bow factory produced a silica-alumina-calcium oxide hard paste body using china clay, and alkaline lime glass cullet, a bone phosphate and a high-magnesian steatitic composite. From these data they correctly concluded that Bow was, in fact, producing a commercially successful, high quality, hard paste body comparable if not exceeding that of the much-esteemed Meissen factory as early as 1743, which predates the first patent of Heylyn and Frye in 1744 and official production of a phosphatic bone china there in 1748 according to the chronological timeline listing for English porcelain production. The Bow porcelain factory actually reached its zenith in the 1750s and thereafter fell into decline until its eventual closure in the 1770s (Ramsey et al. 2001a, b, 2004a, b, 2006).

Although this seems to be a singular triumph for the involvement of analytical science in porcelain studies, when the conclusions are compared with the established historical accounts such as that of Hobson there is an impasse; it has always been maintained that the growth of English porcelain production can be attributed exclusively to the importation of significant technological expertise and information acquired from the major French and German Continental factories such as Meissen and Sevres. Hobson alleged that:

> ... Remember that porcelain was not discovered in England by a process of evolution from the native earthenware. It was, on the contrary, an exotic plant of Eastern origin, naturalized and one might say, hybridized on the Continent, and brought to England, as it were, in cuttings which were planted first in the neighbourhood of London and afterwards disseminated in more congenial soils.

It appears that nothing could be further from the truth and Ramsay & Ramsay are correct in calling for a re-examination of the historical documentation surrounding the Bow manufactory to verify such a statement or to replace it with a sanitised version. How has this situation arisen—have either the analytical or historical data been mis-interpreted? A statement by Ramsay and Ramsay (2007a, b) is relevant to this discussion:

> Central to our work is the 1744 patent of Edward Heylyn and Thomas Frye, two of the Bow proprietors. This patent, sometimes referred to as the Bow first patent in contrast with the bone ash second Bow patent entered by Thomas Frye on 1749, has been misquoted, misunderstood, marginalised and underestimated for over a century.

Citing a host of reputable authors including Chaffers (1863), Church (1881, 1894), Burton (1902), Hobson (1905), Hurlbutt (1926), Tait (1959), Watney (1963, 1973) and Bradshaw (1992).

It is clear that although their stand-alone interpretation is scientifically justifiable, the incorporation of these data and conclusions into the holistic approach whereby the existing historical conclusions are challenged and thence been downgraded to incorrect assumptions from perceived fact has caused sufficient concern to render the analytical data being ignored Yet, one particular aspect of the assertions of Ramsay & Ramsay has not been investigated further historically: namely, they have cited analytical evidence for a particular type of Cherokee china clay called *uneka*, which was imported from the Appalachians in what is now the eastern USA. The unusual chemical composition of this china clay, being 90% halloysite and 10% kaolinite, has facilitated its identification in Bow porcelains dating from 1744 and also several pieces from the so-called A-marked group (Ramsay and Ramsay 2006) which has given a confident assignment of these wares to Bow first period china produced in the period 1743–1745 and therefore probably made by the Bow patentees, Heylyn and Frye. Specimens of this Cherokee clay obtained at source by Ramsay and Ramsay (2007a, b) and analysed by Swenser therein, show 44.8% silica, 38.4% alumina, and minor quantities of titania(0.01%), haematite(0.04%),magnesia (0.03%), lime(0.06%), potash (0.33%) and soda (0.06%): the two major components account for 83.2% of the total mass of *uneka* and the minor components 0.53%—the authors also report a loss in mass of 16.25% upon ignition, this would be attributable to water (or possibly organics) and again indicates an indefinite composition for basing the recipe. The Cherokee clay was described as "*... extremely white, tenacious, and glittering with mica*" by Heylyn and Frye in their 1944 patent and this description has been attested by Ramsay and Ramsay (2006). Church (1881) also comments on the *unaker* china clay source having been among the first to recognise the use of this material in Bow porcelains under the guise of *virgin earth*—he states that in eth second patent taken out by Frye in 1748, the *unaker* is replaced by bone ash:

> The unaker is replaced by other materials. Two parts of virgin-earth, produced by the calcinations of certain animals, vegetables and fossils, are directed to be mixed with one part of flint or sand and fritted; then of this frit two parts are taken and mixed with one part of pipe clay. The glaze was made of red lead, saltpetre and sand with some white lead and smalts. There can be no difficulty in recognising the earth produced by the calcinations of certain animal and vegetable matters with bone-earth, this is calcined bones which consist mainly of phosphate of lime. The patentee, of course, did not desire to be too explicit.

In this transcript, Church does not mention at all the inclusion of glass cullet in the Bow frit but Ramsay and Ramsay (2007a, b) have clearly demonstrated that glass cullet and/or gypsum was an integral component of the Bow paste output.

The work of Ramsay & Ramsay also seriously questions the allegation made by Chaffers (1863) and by Church (1881, 1894) and others that early Bow paste was "unworkable" and that this was deemed to be put in place as a subterfuge to exploit and bolster their patent of 1744. We have already encountered such mystical statements, apparently made with no evidential basis, in relation to Nantgarw and Swansea porcelains made by the same and other authors who have apparently relied on dubious evidential material and perhaps faulty eye-witness accounts which were transcribed incorrectly at the time. These have now been refuted here and elsewhere for Nantgarw and Swansea and perhaps should be seriously examined and reviewed for Bow and some other factories.

9.2 The Impasse Between Analysis and Expert Opinion?

It has frequently been mooted that particularly in the area of easel paintings and watercolours, when the conclusions from scientific analysis agreed with or substantiated those made from the consideration of expert opinion then all was well, but when the two disciplines were providing contrary information the expert opinion always won through. True or not, there are indeed cases where this conclusion has been reached for the most arbitrary of reasons: the opinion of Battie (1994), a well-respected expert on antique porcelains and held in much esteem, cited earlier in this text when answering a query as to the reason analytical science was not used more in ceramics research is a case in point:

> Many tests are invasive-they cannot now, and probably never will, be conducted without damaging the object ... The ultimate test is the expert eye and a collector with perhaps only a few years'experience can still beat any technology available today. And probably always will.

The real reason for the seeming distrust of analytical science in the art arena, therefore, must be related to the interpretations made about the qualitative and quantitative compositions of the porcelain bodies and glazes resulting from the elemental or molecular determinations arising from the compilation of the data. As has been revealed here, the major issue of contention in any analytical determination are:

- The precision of the analytical determinations: this, of course, reflects the sampling regime procedures and the number of replicates taken from each specimen for the analysis. Often, the analytical determinations fail to specify a precise accuracy for the determinations and an assumption of several percent is reasonable: Ramsay and Ramsay (2007a, b) have affirmed that the quantitative elemental determinations obtained during their classic studies of Bow porcelains using scanning electron microscopy and energy dispersive X-ray spectrometry can amount to ±5% or even larger when only small amounts of sample are taken, and Hedges has also stated that the determination of phosphorus in particular is subject to very high errors. Ramsay & Ramsay examined finely ground powders and used a rastering method using a 1 micron probe to achieve bulk compositional data from single point measurements: their internal standards for the determination of elemental occurrences were the following reference materials—apatite (Ca,P), anhydrite (S), plagioclase (Al,Si), tugtupite(Na), Sanidine (K), iron (Fe), lead (Pb), magnesium (Mg) and titanium (Ti). Interference in the determinations was observed for sulfur when lead was present and sodium from fluorescence from the copper sample holder.

- The actual composition of the recipes and formulations for porcelain bodies in the 18th and 19th Centuries, although apparently quite detailed and definitive, is actually rather inexact because of differential sourcing problems. This has already been highlighted in this book with regard to the potential errors arising from the inaccuracies of weighing of the components. The sourcing issues have another imponderable effect: this has been exemplified especially for the china clays and bone ash—and in the latter case there is an additional source of error brought in by the thermal preparation of the source component.
- Finally, the components of the recipes are sometimes not specified with sufficient reliability in the workbooks: for example, Lewis Dillwyn acknowledged that he used glass cullet in his recipe but nowhere is this mentioned in his abbreviated list of components. Occasionally, mention is made of "frit", which can mean almost anything from ground and partially fired clay mixtures with sand to finely ground glass cullet, flints and bone ash. The sourcing of flints would be particularly important in this respect because of inherent impurities such as titania, haematite and gypsum or alabaster. Again, steatite is a magnesium silicate found naturally occurring with other magnesium silicates such as talc. It is also realistic to appreciate that factory owners who are experimenting with porcelain mixtures in efforts to improve their paste bodies would perhaps not have made notes of every modification carried out to their mixture. In this respect a very large unknown is the incorporation of broken, fired shards imported from other porcelain manufactories—note of this has already been made here and excavations at manufactory sites have revealed large dumps of such shards, whose composition when mixed with the recorded recipe would naturally distort the data. Dossie (1758) has recorded the common practice of using ground oriental hard paste porcelain shards mixed with quantities of a flux in the preparation of early ceramics. Godden has recorded an area he found at the Caughley porcelain china works site which contained "a huge quantity of and range of Chinese and English porcelain shards which must have been utilised in the Caughley standard mix".

So, can one be absolutely certain that the porcelain manufacturer actually adhered to his notes critically and observed all the necessary weighings for an accurate assembly of his pastes and glazes? Dillwyn mentions the inclusion of additives such as arsenic oxide, smalt and borax to his porcelain paste but he does not specify the quantities—using the phrase "a little" to describe the addition. We should therefore bear in mind that small differences in porcelain paste composition or of mineral entities determined analytically cannot therefore be ruled to indicate an unambiguous differentiation between the porcelain bodies, inter- and intra-factory. However, the exclusion or non-detection of specific minerals or chemical entities does provide an authoritative discriminatory factor that can be utilised in this way, as has been demonstrated here, and most certainly could be used to expose a fake or a mis-attributed piece masquerading as a genuine rarity from Swansea or Nantgarw.

References

D. Battie, The eyes have it. Antique Collecting J. Antique Collectors Club **28**(9), 3 (March 1994)

P. Bradshaw, *Bow Porcelain Figures circa 1748–1774* (Barrie & Jenkins, London, 1992)

W. Burton, *A History and Description of English Porcelain* (Cassell & Co., London, 1902)

W.B. Chaffers, *Marks and Monograms on Pottery and Porcelain with Historical Notes on Each Manufactory* (1863, J. Davy & Sons, London, Kessinger Legacy Reprints, Kessinger Publishing, Whitefish Montana USA, 2010)

A.H. Church, Cantor Lectures on Some Points of Contact Between the Scientific and Artistic Aspects of Pottery and Porcelain, Lecture IV. *Journal of the Society of the Arts* (1880, January 14), pp. 126–129. (Extended in monograph by Trounce Publishers, London, 1881)

Sir A.H. Church, in *English Porcelain: A Handbook to the China Made in England During the 18th Century as Illustrated by Specimens Chiefly in the National Collection.* A South Kensington Museum Handbook (Chapman & Hall Ltd., London, 1894)

L.W. Dillwyn, *Notes on the Experimental Production of Swansea Porcelain Bodies and Glazes.* Made by Lewis Weston Dillwyn with Samuel Walker at the Swansea China Works Between 1815 and 1817. Presented to the Library of the Victoria &Albert Museum, South Kensington, London by John Campbell in 1920. (Reproduced in Eccles & Rackham, *Analysed Specimens of English Porcelain*, 1922, see reference)

R. Dossie, *The Handmaid to The Arts*, vol II (Published by J. Nourse at The Lamb, opposite Katherine Street, The Strand, London, 1758), p. 342.

R.L. Hobson, *Catalogue of the Collection of English Porcelain in the Department of British and Mediaeval Antiquities and Ethnography of the British Museum* (Printed by the Trustees of the British Museum, London, 1905)

F. Hurlbutt, *Bow Porcelain* (G. Bell & Sons, London, 1926)

W.R.H. Ramsay, A. Gabszewicz, E.G. Ramsay, *Unaker* or Cherokee Clay and its relationship to the bow porcelain manufactory. Trans. Engl. Ceram. Circ. **17**, 474–499 (2001a)

W.R.H. Ramsay, A. Gabszewicz, E.G. Ramsay, The chemistry of A-marked porcelain and its relation to the Edward Heylyn and Thomas Frye patent of 1744. Trans. Engl. Ceram. Circ. **18**, 264–283 (2001b)

W.R.H. Ramsay, J. Hansen, E.G. Ramsay, An A-Marked Covered Porcelain Bowl, Cherokee Clay, and Colonial America's Contribution to the English Porcelain Industry, in *Ceramics in America*, ed. by R. Hunter (Chipstone Foundation, 2004a), pp. 60–77

W.R.H. Ramsay, G. Hill, E.G. Ramsay, Recreation of the 1744 Heylyn and Frye ceramic patent wares using Cherokee Clay: implications for raw materials, Kiln conditions and the earliest english porcelain production. Geoarchaeol. **19**, 635–655 (2004b)

W.R.H. Ramsay, F.A. Davenport, E.G. Ramsay, The 1744 ceramic patent of Heylyn and Frye: unworkable *Unaker* formula or landmark document in the history of english ceramics. Proc. R. Soc. Vic. **118**, 11–34 (2006)

E.G. Ramsay, W.R.H. Ramsay, Bow First Patent Porcelain: New Discoveries in Science and Art, in *The Antiques Magazine* (Brant Publications, New York, September issue, 2006), pp. 122–127

E.G. Ramsay, W.R.H. Ramsay, in *Bow: Britain's Pioneering Porcelain Manufactory of the 18th Century*. The International Ceramics Fair & Seminar, Park Lane Hotel, London, 16 June, 2007a, pp. 1–16

W.R.H. Ramsay, E.G. Ramsay, A Classification of Bow Porcelain from First Patent to Closure: c.1743–1774. Proc. R. Soc. Vic. **119**(1), 1–68 (2007b). ISSN 0035-9211-1-168

H. Tait, *Bow Porcelain 1744–1776: A Special Exhibition of Documentary Material to Commemorate the Bi-Centenary of the Retirement of Thomas Frye, Manager of the Factory and Inventor and first Manufacturer of Porcelain in England* (Published by the Trustees of the British Museum, London, 1959)

B. Watney, *English Blue and White Porcelain of the Eighteenth Century* (Faber & Faber, London, 1963)

B. Watney, *English Blue and White Porcelain of the Eighteenth Century* (Faber & Faber, London, 1973)

Appendix A

European porcelain manufactories, founders and their dates of foundation (Hard paste porcelain black*; Soft paste porcelain red; Black no asterisk unknown paste)

Factory	Year of foundation / closure	Founder
Medici	1575/1587	Francisco de Medici I, Duke of Tuscany
Nevers	1650/1680-1697	Conrade
Rouen	1673/1696	Edme & Louis Poterat
St. Cloud	1693/1766	Pierre Chicaneau/Philippe, Duke of Orleans
Meissen*	1708/1783	Walther von Tschirnhaus / Johann Bottger
Ville L'Eveque	1711/1766	Marie Moreau
Lille	1711/1730	Barthelemy Dorez
Nuremberg	1712/1720	Christopher Marz / Johan Romeli
Vienna*	1718/1864	Claudius du Paquier / Emperor Karl VI
Strasbourg*	1724/1754	Charles-Francois Hannong
Rorstrand*	1726/2005	Johann Wolff
Chantilly	1730/1800	Cirou / Louis Henri de Bourbon, Prince of Conde
Luneville	1731/1749	Duke Francis III
Vincennes	1740/1756	Claude Humbert Gerin
Chelsea	1743/1769	Charles Gouyn / Nicholas Sprimont
Capodimonte*	1743/present	Charles VII, King of Naples and Sicily
Mennecy	1745/1766	Francois Barbin / Duc de Villeroy / Comte d'Eu
Limehouse	1745/1748	Joseph Wilson
Bow	1747/1764	Thomas Frye / Edward Heylyn
Furstenburg*	1747/present	Johann von Langen/Charles I, Duke of Brunswick
Nymphenburg*	1747/present	Joseph Max III, Elector of Bavaria
Villeroy & Boch*	1748/present	Francois Boch
Sceaux	1748/1810	Jacques Chapelle
Longton Hall	1749/1760	William Littler
Vauxhall	1751/1763	Nicholas Crisp / John Sanders
Tournai	1751/1890	Jean Peterinck / Prince Charles of Lorraine
Worcester	1751/present	Dr John Wall / Thomas Flight/Martin Barr
Crepy-en-Valois	1753/1768	Pierre Bourgeois / Louis Gagnefois / Duc de Valois
Orleans	1753/1768	Gerault Darambert / Duc de Penthievre

H. G. M. Edwards, *Nantgarw and Swansea Porcelains*,
https://doi.org/10.1007/978-3-319-77631-6

Frankenthal*	1755/1799	Charles Hannong / Elector Karl Theodor of Bavaria
Sevres	1756/present	Marquis Orry de Fulvy / King Louis XV
Liverpool	1756/1804	Richard Chaffers / Samuel Gilbody / Philip Christian
Limehouse	1757/1831	Joseph Shore
Chelsea-Derby	1757/1848	William Duesbury I / Andrew Planche
Lowestoft	1757/1802	Philip Walker / Obed Alred / Robert Browne
Wedgwood*	1759/present	Josiah Wedgwood
Sitzendorf *	1760/present	Georg Macheleidt / Prince Johann of Schwarzenberg
Berlin*	1763/present	King Frederick II of Prussia
Wallendorf*	1764/present	Johann Wolfgang Hamman / Duke of Saxe -Cobourg
Barancas-Lauraguais	1764/1768	Louis Felicite, Duc de Barncas, Comte Lauraguais
Etiolles*	1766/?	Dominique Monnier / Jean-Baptiste Pelleve
Revol	1768/present	Joseph-Marie & Francois Revol
Plymouth/Bristol*	1768/1770	William Cookworthy / Richard Lund
Niderviller*	1769/1827	Paul Cyfflie / Duc de Lorraine
Bristol*	1770/1781	Richard Champion / William Cookworthy
Chelsea-Derby	1770/1784	William Duesbury I
Arras	1770/1790	Boussemot / Delhaye
Limoges*	1771/present	Turgot / Comte de Artois / King Louis XVI
Bourg-de-la-Reine	1773/1804	Joseoh Juililen
Clignancourt*	1775/?	Deruelle / "Monsieur", Brother of King Louis XVIII
Caughley	1775/1799	Ambrose Gallimore / Thomas Turner
Royal Copenhagen*	1775/present	Frantz Heinrich Mueller / King Christian VII
Spode	1776/present	Josiah Spode I
New Hall*	1781/1835	Messrs Hollins, Keeling , Warburton, Clowes
Rabbensgrun	1789/1793	Franz Haberditzei
Schlaggenwald	1793/?	Johann Paulus / Georg Reumann/Louis Grener
Klosterle	1794/1945	Nicholaus Weber / Johann Sonntag / Count Thun
Davenport	1794/1887	John Davenport
Ridgway	1794/present	Job & George Ridgway
Coalport	1795/1841/present	John Rose
Pinxton	1796/1806	John Coke/William Billingsley
Swansea	1812/1822	Lewis Dillwyn / William Billingsley
Nantgarw	1817/1822	William Billingsley / William Weston Young

Appendix B

Document 1: ***Notes on the Experimental Production of Swansea Porcelain Bodies and Glazes Made by Lewis Weston Dillwyn with Samuel Walker at the Swansea China Works between 1815 and 1817.*** **Presented to the Library of the V&A Museum by John Campbell in 1920. Taken with modification from Eccles & Rackham,** ***Analysed Specimens of English Porcelain*****, 1922.**

***Key to abbreviations in the original text*:**
V sand; KO flint; LO lime; YX bone; B St. Stephen's clay; E Norden clay; FO composition, china stone, FX pearl ash; EX nitre; GX arsenic; AX lead; MX borax; DX glass; LX smalt; SR soaprock; No. 157 sand frit.

Body Number 1: Porcelain body.
12V sand + 1 FX pearl ash—fine.
10 FO china stone + 1 FX pearl ash—coarse.
No glass content.

Body Number 2: Glass frit: 11 V sand + 9 FO china stone + 6 FX pearl ash + 3 MX borax - 26 parts taken with 12 AX lead + 1 SR soaprock.
The first mention of a glass frit with high lead oxide content.

Body Number 3: 3 V sand + 3 FO china stone + 2 FX pearl ash, fritted with one-tenth of SR soaprock.
The above is a "Variation from the Nantgarw body". (*Author's note*: *Nantgarw porcelain does not contain any glass frit or lead glass component—is this what Dillwyn means by a variation from Swansea in the Nantgarw body and does this imply that he knew the Nantgarw porcelain formulation?*).

Body Number 4: Common porcelain body.
12 FO composition china stone—4cwt 70lb; 8 YX bone ash - 5 bone ash; 8 B St. Stephen's china clay—3½; 1 E Norden blue clay—35 lb.

H. G. M. Edwards, *Nantgarw and Swansea Porcelains*,
https://doi.org/10.1007/978-3-319-77631-6

Note: no sand used. The quantities of paste mixture specified seem large for a trial experiment—some 518 lbs of china stone alone (equivalent to 234 k). (According to Eccles & Rackham, this is the "*first evolution*" from a duck-egg porcelain body).

Body Number 5: Biscuit recipe first used in autumn 1815.
20 parts V sand + 1 part FX pearl ash in water fritted at very high heat.
140 lbs of above frit + 110 FO china stone + 25 SR soap rock fired very regularly and gradually or it will blister. It was afterwards found that the blistering proceeds from an accidental mixture of alabaster to prevent the possibility of which great care must be taken.
(Believed to be an early effort at a *trident* body).

Body Number 6: Variation on the above formulation. Sounder body but the articles still continue to "fly with hot water".
140 lb frit + 11O lb FO china stone + 35 lb SR soap rock.
It was discovered that the B St. Stephen's clay and FO china stone when fritted together into one mass with FX pearl ash make an equally good looking body which stands well.

Body Number 7: 45 FO china stone composition + 10 LO lime + 28 B St. Stephen's china clay.
Makes a body which comes very near the Chinese eggshell and will take a hard glaze but must be fired at a very high temperature and pieces are then apt to get out of shape.

Body Number 8: Very good.
9 parts V sand + 1 part B St. Stephen's clay and a little LO lime fritted in a very high heat.
3 of the above frit + 3 FO china stone + 1/10 SR soap rock.

Body Number 9: B St. Stephen's clay of which half has been fritted and ground, glazed with FO china stone, is the *Dresden china*. A very great heat is necessary and difficult to get saggars to stand it. Equal parts of B St. Stephen's clay and FO china stone is the very best French china and will take an FO china stone glaze. No other than an FO china stone glaze will do as all others craze.

Body Number 10: A beautiful china which stands well but is rather too soft for the hard glaze.
12 B St. Stephen's china clay + 12 YX bone ash + 12 FO china stone + 3 LO lime.
(According to Eccles & Rackham, page 15, this approximates very closely to the Nantgarw body recipe). No glass content.
Autumn 1816.

Body Number 11: A beautiful body and in all respects answers.
3 B St. Stephen's clay + 3 FO china stone + 3 YX bone ash.

Body Number 12: An improvement.
8 B St. Stephen's china clay + 7 FO china stone + 8 YX bone ash.
(According to Eccles & Rackham this is a *second evolution* on the duck's egg body).

Body Number 13: Makes the body harder but large pieces are more apt to fly.
9 B St. Stephen's clay + 9 YX bone ash + 7 FO china stone.
This body glazes well with glaze number 2.
March 1817.

Body Number 14: 12 V sand + 10 FO china stone + 2 FX pearl ash fritted together then 14 of this frit + 2 SR soap rock makes a beautiful and good body. If only 1 SR soap sock is used it makes the body whiter but the clay is more difficult to work. Afterwards, the following alteration was made.
(A second attempt at the *trident* body).

Body Number 15: Without much improvement on the above body 8 V sand + 6 FO china stone + 1 FX pearl ash fritted in a high heat which had better exceed the biscuit heat.
This body glazes well with glaze number 1.
December 1817.

Body Number 16: Makes a beautiful white opaque body and with glaze number 3 is the finest earthenware I ever saw.
24 YX bone ash + 8 KO flint + 16 B St. Stephen's china clay + 5 E Norden clay + 1 LX smalt.

Glaze Number 1: Frit: 10 FO china stone + 6 LO lime + 2 B St. Stephen's clay + 12 V sand + 14 AX lead + 8 MX borax calcined + 4 EX nitre : total 56 parts.
Run in the glaze kiln or earthenware biscuit kiln which is about the same heat I prefer the latter on account of its longer continuance, which makes the frit run more thoroughly throughout.
Then 56 parts of the above frit + 30 FO china stone + 6 LO lime + 2 B St. Stephen's clay + 14 AX lead + ½ GX arsenic.

Glaze Number 2: 24 V sand + 12 LO lime + 6 AX lead + 16 MX calcined borax + 2 FX pearl ash.
Run in glaze heat or as Number 1.
Then 28 parts of the above frit + 40 FO china stone + 28 AX lead + 6 LO lime + 4 B St. Stephen's clay.

Glaze Number 3: 24 V sand + 12 LO lime + 6 AX lead + 16 MX borax calcined + 2 FX pearl ash frit in glaze heat as Number 1; then 48 FO china stone + 6 LO lime + 4 B china clay + 30 AX lead + 40 above frit + ½ arsenic to be dipped thick.

Appendix C

Mohs scale of hardness for minerals relevant to porcelain manufacture

Mohs number	Mineral	Formula	Absolute hardness
1	Talc	$Mg_3SiO_{10}(OH)_2$	1
	Soapstone	$Al_2Mg_3(SiO_3)_4$	
1.5	Graphite	C	2
2	Gypsum	$CaSO_4 \cdot 2H_2O$	3
	Kaolin	$Al_2Si_2O_5\ (OH)_4$	
2.5	Rock salt	NaCl	6
	Borax	$Na_2B_4O_7 \cdot 10H_2O$	
3	Calcite	$CaCO_3$	9
4	Fluorite	CaF_2	21
5	Apatite	$Ca_5(PO_4)_3(OH^-, Cl^-, F^-)$	48
5.5	Glass	SiO_2	60
	Haematite	Fe_2O_3	
	Smalt	$CoAl_2O_4$	
6	Orthoclase feldspar	$KAlSi_3O_8$	72
	Anatase	TiO_2	
7	Quartz	SiO_2	100
	Unglazed porcelain	frit	
	Sand, flint	SiO_2	
7.5	Spinel	$MgAl_2O_4$	150
8	Topaz	$Al_2SiO_4(OH^-, F^-)$	200
8.5	Silicon nitride	SiN	300
9	Corundum	Al_2O_3	400
9.5	Carborundum	SiC	950
10	Diamond	C	1500

H. G. M. Edwards, *Nantgarw and Swansea Porcelains*,
https://doi.org/10.1007/978-3-319-77631-6

Index

H. G. M. Edwards, *Nantgarw and Swansea Porcelains*,
https://doi.org/10.1007/978-3-319-77631-6

The manufacturer's authorised representative in the EU is Springer Nature Customer Service Centre GmbH, Europaplatz 3, 69115 Heidelberg, Germany. If you have any concerns regarding our products, please contact ProductSafety@springernature.com

Printed and bound by CPI Group (UK) Ltd, Croydon, CR0 4YY
05/10/2026
02228144-0002